Resource Management on Distributed Systems

IEEE Press
445 Hoes Lane
Piscataway, NJ 08854

Resource Management on Distributed Systems

Principles and Techniques

Shikharesh Majumdar
Department of Systems & Computer Engineering
Carleton University
Ottawa, Canada

IEEE Press
WILEY

Published by John Wiley & Sons, Inc., Hoboken, New Jersey.
Published simultaneously in Canada.

Library of Congress Cataloging-in-Publication Data applied for:
Hardback ISBN: 9781119912934

Cover Design: Wiley
Cover Image: © metamorworks/Shutterstock

Set in 9.5/12.5pt STIXTwoText by Straive, Chennai, India

To my teachers and students, who continuously refreshed my knowledge pipeline.

Contents

About the Author

Shikharesh Majumdar is a Chancellor's Professor and the Director of the Real Time and Distributed Systems Research Centre at the Department of Systems and Computer Engineering in Carleton University in Ottawa, Canada. He holds a Ph.D. degree in Computational Science from University of Saskatchewan, Saskatoon, Canada. His research interests are in the areas of cloud and grid computing (with the current focus on Big Data processing), smart systems, operating systems and performance evaluation. Dr. Majumdar has published extensively in these areas. He actively collaborates with the industrial sector and has been the area editor for the Simulation Modelling Practice and Theory journal published by Elsevier. Dr. Majumdar is a Fellow of the Institution of Engineering and Technology (FIET), a member of ACM, a life senior member of IEEE and a Professional Engineer (Ontario, Canada). He has received multiple international awards and was a Distinguished Visitor for the IEEE Computer Society.

Preface

The availability of processors, memory, and high-speed interconnection networks at a reasonable cost is continuously increasing the use of parallel and distributed systems. Appropriate management of resources is crucial, however, for effectively harnessing the power of the underlying resource pool. While resource management techniques on conventional single processor systems is covered in many standard operating systems books, there is comparatively less coverage of resource management on parallel and distributed systems in the currently existing books. This book aims at addressing this gap and describing algorithms and techniques for managing resources on parallel and distributed systems, including grids, clouds, smart facilities, and parallel processing-based platforms for big data analytics.

The book focuses on resource management on distributed systems, which is of interest to students, researchers, and industrial practitioners who work with systems comprising multiple resources that may range from clusters to clouds to smart facilities. In addition to a discussion of existing knowledge in the area, the book includes material based on research results that have made significant contributions to the state of the art. The book describes key concepts as well as summarizes research results. The key features of the book include the following:

- An introduction to five general principles of resource management that will be adapted by the techniques described in the following chapters.
- A description of the resource management techniques and algorithms with a discussion of their impact on system performance, energy usage, and cost. When applicable, attention is paid to the trade-offs among these three characteristics. Special techniques for achieving system scalability are discussed.
- Results of applying the techniques on simulated as well as real systems (including clouds and big data processing platforms).

- In addition to data processing on cloud environments topics that include the following are discussed:
  - o Big data platforms and frameworks, e.g. MapReduce.
  - o Sensor-based smart systems that are becoming an important component of a smart society.
- Research results that include the description of the different experiments and pointers to research papers that provide supporting documentation for the research described. Insights into system behavior and performance resulting from these research results are provided.

The organization of the book is influenced by many years of teaching graduate courses in distributed systems in general and resource management in particular, as well as my research performed in the area. Examples and exercises are included in appropriate sections of the book.

Book Contents

This book concerns resource management in distributed systems that can be classified into two categories: computing-intensive systems and data-intensive systems. The book has two parts, each focusing on a particular type of distributed systems. Part 1 (Chapters 2–5) focuses on issues underlying distributed computing-intensive systems, whereas Part 2 (Chapters 6–10) is concerned with distributed data-intensive systems.

Basics: Introductory material, including definitions of basic units of distributed computations and their characteristics, are discussed in Chapters 1 and 2.

Chapter 1 describes the evolution of distributed systems from nodes communicating with one another using basic communication mechanisms such as remote procedure calls (RPCs) to clusters, grids, and clouds, including edge computing systems and smart facilities. The basic units of computation used by various applications such as threads and processes are introduced. Three types of resource management operations performed on these computation units that include allocation and sheduling are described. General principles underlying resource management that form the backbone of a number of resource management techniques described in the later chapters are introduced.

Chapter 2 focuses on the characterization of parallelism in applications. Both graph-based characteristics as well as single-point characteristics are introduced. These application attributes have an intimate relationship with the execution behavior and performance of the system running the application and are thus important in the context of resource management. Performance metrics that can be used for analyzing the performance of resource management algorithms are described in this chapter. Energy consumed by computation and data-intensive applications is often of critical concern. The later part of the chapter introduces energy-related metrics and characteristics and discusses their interrelationship.

The trade-off between energy consumption and performance is discussed. Energy-aware resource management is the subject of discussion in a later chapter.

Allocation and Scheduling: Resource allocation and scheduling are two important resource management operations that are discussed in Chapter 3. The resource allocator maps the application work units to processing resources and determines which process/thread will be executing on which processor. Well-known results in the area are discussed. The chapter discusses both optimal algorithms as well as techniques for devising heuristic resource allocation techniques. The discussion of resource allocation is followed by a discussion on process/task scheduling. The task scheduler decides the order in which computation units allocated to a processor will execute. Scheduling of tasks for an application with service level agreements that include deadlines for completion is considered. Scheduling algorithms for both single processor systems as well as systems with multiple processors are described. Analyses of performance of these algorithms are presented. The chapter ends with a discussion of scheduling techniques for client server systems.

Handling uncertainties in allocation and scheduling: Real systems are often characterized by uncertainties associated with system and workload parameters. Chapter 4 focuses on systems with such uncertainties and describes how to build in robustness into resource management techniques to mitigate their adverse impact on system performance. Two types of uncertainties are discussed. The first results from the errors associated with user-predicted task execution times that are often specified as part of a service level agreement (SLA). Techniques for handling both underestimation and overestimation of task execution times are discussed. Analyses of performance of the techniques are presented. A cloud data center often comprises hundreds and thousands of computing resources that are susceptible to changes with time and the exact local scheduling policy used by each resource may not always be available a priori to the resource manager for the data center. Techniques for resource management to handle this second type of uncertainty associated with the knowledge of the local scheduling policies used at the various resources are described. Performance analyses of these techniques referred to as techniques for resource management "in the dark" are presented.

Handling changes in workload in resource allocation: Determining the number of resources to provision for a given workload is a complex undertaking. The problem is further complicated by dynamic changes in workload that are typically the case on a distributed system used by multiple users. Chapter 5 addresses the problem of dynamically controlling the number of CPU resources by using resource auto-scaling. System capacity is increased or decreased automatically by the auto-scaling algorithm so that client satisfaction is met while keeping the cost of resource usage under control. The concepts of both vertical auto-scaling in which the CPU power and memory capacity of a given resource are controlled in accordance with the system load, and horizontal auto-scaling that increases or decreases the number of computing resources in accordance with the change in system workload are introduced. Most of the chapter concerns horizontal auto-scaling techniques. Three types of horizontal auto-scaling are discussed: (i) reactive auto-scaling for which a change to the number of resources is made after a change has occurred to the system workload; (ii) proactive auto-scaling for which the future system workload is predicted and the change in the number of resources to handle a change in this future workload is computed proactively before the said change in workload intensity occurs; and (iii) hybrid auto-scaling that is a combination of reactive and proactive auto-scaling. Performance analysis for each technique is reported.

Data-Intensive Systems and MapReduce: Part 2 of the book that concerns data-intensive distributed systems starts with Chapter 6, which focuses on platforms running MapReduce jobs that are used in big data analytics as well as for other data-intensive applications. The chapter describes techniques for allocation and scheduling for MapReduce jobs associated with SLAs that include job completion deadlines. Two resource management algorithms, a budget-based algorithm and a constraint programming-based algorithm, are discussed. The SLA associated with a job includes user estimates of task execution times that are often subject to error. Two techniques for handling such errors and increasing the robustness of resource management are described. The chapter includes a thorough discussion of the performance of the various techniques described.

Energy-Aware Resource Management: Both compute-intensive as well as data-intensive applications can lead to increased energy consumption of servers as well the air conditioning equipment used for cooling the servers. The cost for energy consumption is a significant component of the total operational cost of a data center. A high energy consumption often leads to higher carbon dioxide emissions. Controlling the energy consumption is thus an important goal for operators of cloud data centers. There is a trade-off, however, between performance and energy consumption. Energy-aware resource management techniques for clouds subjected to MapReduce jobs are the subject of attention in Chapter 7. The chapter presents algorithms that attempt to reduce energy consumption while meeting the SLA requirements of the system as well as a thorough performance analysis of these techniques.

Streaming Data: Analysis of big data requires the ability to handle both stored and streaming data. The two previous chapters focused on resource management techniques for stored data. Chapter 8 is concerned with streaming data, which is characterized by a continuous stream of data tuples that are often referred to as messages. A number of platforms for analyzing streaming data are described. The first part of the chapter concerns dynamic priority scheduling that can support the dynamic changing of priority of applications based on data contents. The second part of the chapter describes systems used for complex event processing. Complex event processing (CEP) is used to handle real-time issues in the context of big data analytics. It facilitates the analysis of multiple streams of data in real time to generate alerts or indicate possible opportunities to be exploited on the system. The CEP concept is explained first, followed by the description of a mobile, smart phone-based CEP system for analyzing the data collected by sensors attached to the body of a patient.

Data Management: In addition to processing and storage devices, data is considered an important resource for enterprises, scientists, and engineers. Management of this data resource is the subject of discussion in Chapter 9. Two types of data management techniques, indexing and filtering, which can speed up the processing of data, are considered. Indexing reduces the latency of performing searches on a document set with a large volume. An index is used to identify the list of documents that contain a specific keyword. Both the well-known inverted indexing technique and a newer graph-based indexing technique are described. Their usefulness in the context of both simple and complex queries is discussed. The performance analysis of these techniques is presented. The second part of the chapter focuses on data filtering. Often, the user is interested in a certain subset of data characterized by specific keywords.

Data filtering techniques use the user-specified keywords for filtering out the data required by the user from a large volume of raw data. Only the filtered data is stored for future search by the user's applications. Since the data volume is significantly reduced, the latency associated with searches is much smaller than that achieved when the whole raw data is searched. The filtering operation in itself is time consuming and parallel processing based filtering techniques are described. Both filtering algorithms for stored and streaming data, as well as their performance analysis, are described in this chapter.

Sensor-Based Systems: Chapter 10 continues the discussion of data-intensive systems. It concerns resource management in sensor-based systems and smart facilities that often leverage sensors for their monitoring, control, and management. Various types of smart facilities such as sensor-equipped bridges and research collaboration platforms are described. Middleware services that are used in the management of smart facilities are introduced first. The architecture of a sensor-based bridge management system and a cloud-based collaboration platform for researchers in smart facilities are described. Wireless sensor networks are often used in smart facilities. Resource management on wireless sensor networks shared by multiple applications is discussed in detail. Both techniques for sensor allocation and sensor request scheduling are described. In addition to a description of resource management algorithms, the chapter includes performance analysis of the different allocation and scheduling algorithms.

The book is concluded in Chapter 11, which provides a summary of the material presented in the different chapters.

Acknowledgments

Writing this book turned out to be a long and arduous process. Thanks to Suparna for instilling the book writing idea in me and for her continuous support in the past few years during which the book was written. I am indebted to my late mother, Stuti, and father, Biresh, for teaching me the value of education and innovation that have inspired me to follow a research-based career later in my life.

A significant part of the book content is based on my research at Carleton University in Ottawa. I would like to thank all my graduate students, as well as both my industrial and academic collaborators, who have contributed to my various research projects. I am grateful to Carleton University for their support in facilitating my research work. The support provided by the staff at Wiley and IEEE Press is highly appreciated.

1

Introduction

1.1 Introduction to Distributed and Parallel Computing

A rapid development in computing and communication technology over the past few decades has enabled the building of today's distributed processing systems. The invention of microprocessors in the 1970s and the subsequent large improvement in clock speed every few years have led to their capability to process billions of instructions per second. The enhancement in VLSI technology has made it possible to pack a very large number of transistors in a small area on silicon. This, in turn, has brought in the era of multicore processors that are used in most computers of today. The development in processor technology is also accompanied by a tremendous improvement in network technology that has given rise to a few orders of magnitude in network bandwidth over the last few decades. The availability of high-speed processors at a reasonable cost and the ability to connect them with high-speed networks are leading to the proliferation of distributed and parallel systems of today including grids and clouds.

More recently, the developments in wireless technology have been changing the face of distributed systems. Wireless systems such as smartphones and computers equipped with wireless adaptors are now becoming components of distributed systems. The availability of sensors at low cost and the advent of Internet of Things (IoT) are fueling the incorporation of distributed system technology in cyber–physical systems that include smart buildings, autonomous vehicles, and smart health monitoring systems.

The evolution of distributed systems from a mere collection of a small number of processes sharing computing resources to clouds that include hundreds or thousands of servers and smart facilities is the subject of discussion in the following sections of this chapter.

Resource Management on Distributed Systems: Principles and Techniques,
First Edition. Shikharesh Majumdar.

1.2 Types of Computing Environments

A computing environment provides the hardware and software infrastructure for the execution of one or more applications. Three broad categories of such systems are described.

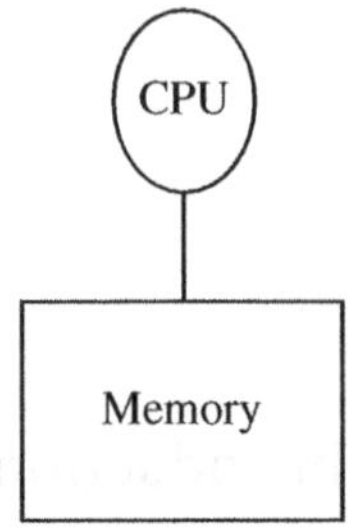

Figure 1.1 Single-processor-based system.

Single-Processor-Based System: Such systems are characterized by a single CPU and memory sub-system (see Figure 1.1). A program running on the system uses the CPU and memory. When multiple independent programs are run on the system at the same time the system is referred to as a *multiprogrammed* system.

Shared Memory (Multiprocessor) Systems: These systems deploy multiple processors that can run components of a given application (see Figure 1.2). These components may communicate with each other through shared variables that are stored in the shared memory or through message passing. When components of multiple programs are active simultaneously on the system it is referred to as a multiprogrammed parallel system (Majumdar et al. May 1988).

Distributed Memory Systems: Such a system is characterized by multiple sets of CPUs each equipped with its own local memory connected to each other through an intercommunication network (see Figure 1.3). Multiple components of a given application can run on multiple processors and communicate with each other through messages.

Since there is no shared memory explicit message passing is used for intercommunication among the program components. As in the case of the other two types of systems a distributed memory system can either run a single application or be multiprogrammed supporting the execution of multiple applications at the same time.

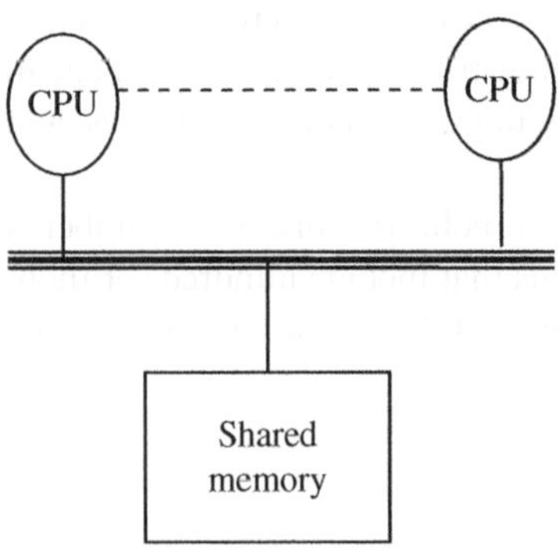

Figure 1.2 Shared memory system.

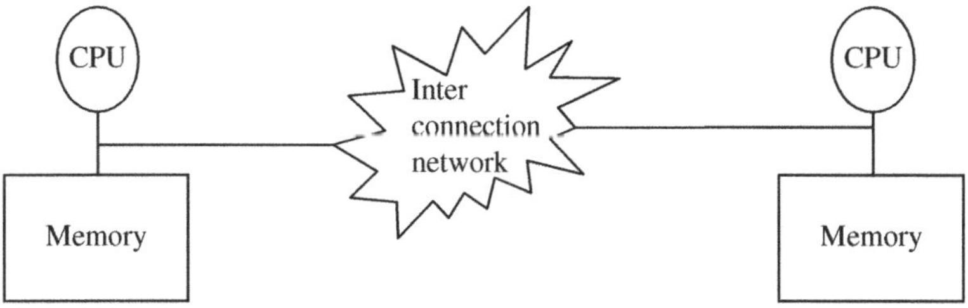

Figure 1.3 Distributed memory system.

1.3 Units of Computation

A computation may comprise a number of components that are often referred to as processes and threads. Both processes and threads are schedulable entities that can compete for execution on the system. Each of these is briefly described.

1.3.1 Process

A process is a program in execution (Silberschatz et al. 2014). A process is characterized by the application code that it executes, the memory it uses as well as a state. The memory used by a process includes a data section for the storage of variables, a stack used, for example, for passing parameters to the procedures/methods that are called as well as a heap or memory that is allocated dynamically to the process during its execution.

A process is characterized by a state that describes what it is doing at a given point in time. A process makes a transition from one state to another based on an event that has occurred in the system. Different characterization of process states are used in the literature. The most common ones include three states that are described next.

Running: state during which the process is executing instructions on a CPU.

Waiting/Blocked: state during which the process is not in execution and is waiting for an event to occur or for a request such as a request for performing Input/Output (I/O) to be completed.

Ready: state during which the process is ready for execution but not running on the CPU yet. A simple state transition diagram characterized by these three states is presented in Figure 1.4. A transition from ready to running is made when the operating system decides to *schedule* the process on the CPU. While executing the process may *request* for performing an I/O or wait for the occurrence of an event and enter the waiting (blocked state). The process makes a transition back to the ready state when the operation it is waiting for is completed. Note that a

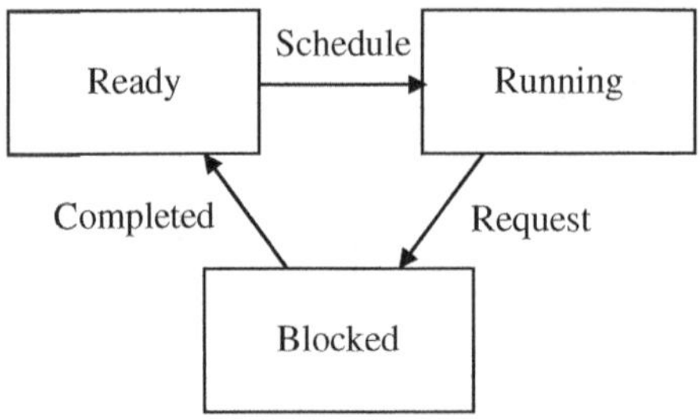

Figure 1.4 State transition diagram for a process.

more elaborate characterization of process states that include additional states is available in the literature (see Silberschatz et al. 2014 for example).

Operating systems maintain a data structure called the *process control block (PCB)* for managing processes. The components of this data structure vary from one operating system to another. The key elements of a PCB that are maintained by most operating systems include a Process Id (PID) that uniquely identifies a process, the process state, fields for storing the CPU register contents when the process is interrupted, scheduling information such as process priority, accounting information including the amount of system resources consumed by the process and information regarding the files and I/O used by the process.

1.3.2 Threads

A process (described in the previous section), often referred to as a heavyweight process, may comprise multiple threads, also referred to as lightweight processes. All the threads in a process may run concurrently on multiple cores or multiple processors. On a single-core CPU, when the currently executing thread in the process blocks another ready thread from the same process may be allowed to run on the CPU. Each thread is associated with a stack, a program counter and register set. It typically shares the global data set and code with the other sibling threads in the same address space corresponding to the parent (heavy weight) process.

With a multithreaded operating system kernel there can be threads both at the user level and kernel level. A user-level thread needs to be mapped to a kernel-level thread before it can execute. Multiple threading models exist. With a one-to-one model, each thread is associated with a single kernel thread that is used when the respective user thread executes. In a many-to-one model, many user-level threads can be mapped to the same kernel thread. With a many-to-many model, many user threads can be mapped to multiple kernel threads. Thus, a given user thread may run on different kernel threads when it is invoked multiple times. Further discussion of the threading models is not important in the context of this book. The interested reader may find such a discussion in standard books on operating systems.

1.3.3 Resource Management Operations

Three key resource management operations are described in this chapter: allocation, scheduling and system provisioning that include the determination of the number of resources to be deployed. Each of them is briefly described.

Allocation: Given a number of processors and a number of processes or threads the allocation operation refers to the mapping of which process (thread) is to be executed on which processor.

Scheduling: refers to the order in which processes (threads) are to be executed. Given a number of processes allocated to a given processor the scheduler determines the time at which each process will be executed.

Note that allocation and scheduling operations can be performed separately but in the case of some techniques both of these operations are performed at the same time.

Provisioning: includes the determination of the number of processors needed to be used for handling a given workload.

1.4 Principles Underlying Resource Management

Parallel and distributed computing makes multiple resources available to its users. The advantages of distributed systems include concurrent execution of software, system reliability, and the unification of distributed resources. Management of resources is of utmost importance for the effective use of distributed systems and is the focus of attention of this book. Resource management techniques described in various chapters of this book are often based on a number of general principles that include the following.

- *Principle 1: Use Knowledge of Workload Characteristics*
- *Principle 2: Monitor and Adjust*
- *Principle 3: Use Knowledge of System Characteristics*
- *Principle 4: Perform Load Balancing*
- *Principle 5: Static versus Dynamic*
  - *React to changes in systems state*
  - *Minimize overheads*

1.4.1 Principle 1: Use Knowledge of Application/Workload Characteristics

This principle refers to using the knowledge of application characteristics, such as the execution time of a process. Using *a priori* knowledge of process execution time is discussed in the literature in the context of scheduling on single

CPU-based multiprogrammed systems. Examples include shortest job first (SJF) in which a process with a shorter execution time is given priority over processes with a higher execution time. Preemptive version of SJF known as the shortest remaining time first (SRTF) is known to be the optimal CPU scheduling algorithm on single CPU-based multiprogrammed systems (Silberschatz et al. 2014).

Knowledge of the characteristics of the workload the system is subjected to is often useful in resource and system management. The distribution of the inter-arrival times of jobs is known to have an intimate relationship with the performance of a queuing system (Kleinrock 1975).

On a conventional single CPU-based multiprogrammed system, a good balance is to be made between the number of CPU-bound jobs (that are characterized by long CPU bursts and a small number of I/O operations) and I/O-bound jobs (that are characterized by short CPU bursts and a large number of I/O operations). The resource manager tries to maintain a balanced multiprogramming mix such that good utilization is achieved for both the processing and I/O resources.

One of the problems with algorithms such as SJF and SRTF is that they need exact *a priori* knowledge of application characteristics such as process execution time. The characteristics of the workload that are expected to arrive in the future are often unknown.

The past history of system behavior, particularly the characteristics of the jobs or applications that have been already executed on the system can be used to predict the characteristics of the jobs that arrive on the system in the near future. Two such techniques are discussed.

(i) *Exponential Averaging:* is a technique that is used to predict the characteristics of a job such as its next CPU burst time. If TP_n is the predicted duration of the nth CPU burst duration and TA_n is the actual duration of the nth CPU burst then the following equation can be used to predict the value of the nth CPU burst time:

$$TP_n = \alpha\, TA_{n-1} + (1 - \alpha)\, TP_{n-1} \tag{1.1}$$

where $0 \leq \alpha \leq 1$.

When predicting the value of the nth CPU burst duration the previous CPU burst duration TA_{n-1} as well the predicted value TP_{n-1} computed in the last iteration are already known and can be used to predict the duration of the nth CPU burst. The right-hand side of Eq. (1.1) aptly captures the previous history of CPU burst durations. The first term corresponds to the immediate history (the previous CPU burst) whereas the second captures the history of all previous burst durations. This follows from the expansion of the TP values computed from Eq. (1.1)

$$TP_{n-1} = \alpha\, TA_{n-2} + (1 - \alpha)\, TP_{n-2} \tag{1.2}$$

$$TP_{n-2} = \alpha\, TA_{n-3} + (1 - \alpha)\, TP_{n-3} \tag{1.3}$$

$$\ldots$$

Equation (1.2) can be used to substitute TP_{n-1}, Eq. (1.3) can be used to substitute $TP_{n-2,}$ and so on. All the values on the right-hand side of the previous equations are known or can be computed from other known values. Note that the value of α determines the weight attached to immediate and past CPU burst characteristics that occurred on the system.

(ii) *Machine-Learning-Based Prediction*: Predictive machine learning techniques have been used to predict the number of resources to be acquired in the future in an auto-scaling system that adjusts the number of resources in the client's cloud based on the predicted workload in such a way that the desired quality of service (QoS) can be met. The workload is predicted based on the past history of the workload that has already executed on the system. If for example the predicted characteristics of client requests are predicted lead to a large number of SLA violations in the near future the system decides to increase the number of resources whereas if on the other hand the predicted workload is deemed to give rise to idle resources on the system the auto-scaling algorithm, decides to reduce the number of resources such that a cost saving for the client is achieved.

1.4.2 Principle 2: Monitor and Adjust

When prior knowledge of applications and or workload is not available the system can monitor the system behavior during run time and perform resource management. Two examples: Multilevel Feedback Queue (MLFBQ) and Reactive Auto-scaling are discussed next. The first is concerned with CPU scheduling for jobs whereas the second focuses on auto-scaled systems.

MLFBQ (Silberschatz et al. 2014): is a CPU scheduling policy that organizes processes in multiple levels of priority queues. Each queue level is associated with a time slice that is used by the processes that are currently queued in that level. If a process does not complete its CPU burst before the end of the time slice associated with its queue level, it is demoted in priority, and it joins the lower-level priority queue. This way processes with long CPU bursts eventually execute at a lower priority level, while processes with short CPU bursts finish their CPU burst at a higher priority level. This is a heuristic strategy that tries to mimic the behavior of the well-known Shortest Remaining Processing Time (SRPT) scheduling policy without using exact a priori knowledge of CPU burst characteristics of jobs. It just monitors the execution behavior of current jobs and adjusts their priorities accordingly. It attempts to run jobs with higher CPU burst times with lower priority.

Reactive Auto-scaling: is a technique that can be used on a cluster or cloud subjected to job requests with SLAs. In such a system, a request needs to be rejected if the respective SLA cannot be met. Auto-scaling on these systems is used to dynamically adjust the number of resources on the system such that a specific Grade of

Service (GoS) can be maintained. An example of a GoS is the proportion of requests that are accepted by the system. One way to improve the acceptance ratio is to acquire more resources, that is, however, accompanied by an increase in cost for the user. A reactive auto-scaling system monitors the acceptance ratio and adjusts the number of resources accordingly. It increases the number of resources if the acceptance ratio falls to a level that is too close to the GoS and decreases the number of resources if the acceptance ratio is far above the specified GoS.

1.4.3 Principle 3: Use Knowledge of System Characteristics

System characteristics include attributes of the system such as CPU speed, memory capacity, network bandwidth, number of CPUs on the system as well as software characteristics such as the local operating system used by a node in the distributed system. In case there are multiple machines the characteristics of each machine in terms of the attributes listed in the previous sentence can be useful when allocating resources or scheduling tasks on the system.

Sometimes, the use of both system and application characteristics can be useful in resource management. For example, if two processes intercommunicate heavily and the interconnection network connecting multiple CPUs is slow it may be better to allocate both processes on the same CPU.

1.4.4 Principle 4: Perform Load Balancing

On systems using multiple processing elements, it is important to balance the distribution of work on these elements. Consider, for example, a batch of three jobs with execution times of Job 1, Job 2, and Job 3 being 60, 120, and 180 minutes, respectively. The time at which all the jobs in the batch are complete, referred to as the batch completion time, is often the performance metric of interest. A balanced allocation on a system comprising two identical CPUs leads to the allocation of Job 1 and Job 2 on CPU 1 and Job 3 on CPU 2. This will lead to a batch completion time of 180 minutes. An unbalanced allocation that leads to the running of Job 1 on CPU 1 and Job 2 and Job 3 on CPU 2 that will give rise to an inferior performance as the batch completion time will increase to 300 minutes.

1.4.5 Principle 5: Static Versus Dynamic

Both static and dynamic resource management are possible. For the first, the processes allocated to a processor or the schedule of execution of processes on a processor that executes multiple processes do not change, whereas for dynamic resource management scheduling and allocation of processes to processors can change dynamically with time. Static techniques often exhibit a smaller resource management overhead in comparison to dynamic resource management. This is because static resource allocation and scheduling are performed only once whereas the number of resource management operations is likely to be more in the case of dynamic techniques since resource management is performed multiple times.

In addition to allocation and scheduling determining the number of processors can also be performed statically or dynamically. For example, on a cluster, the number of processing resources may be determined during system provisioning and held unchanged during the lifetime of the system. In comparison to using this static provisioning approach, the system can be auto-scaled, and the number of processors on the system can be changed dynamically based on system load, for example.

The choice between static and dynamic resource management needs to be performed carefully and dynamic resource management is justifiable only when the overhead in performing dynamic resource management is offset by the concomitant improvement in system performance.

1.5 Evolution of Distributed Systems

Many forms of distributed systems are being used today. Distributed computing has evolved in the past few decades from computing performed on a mere collection of communicating nodes to feature-rich clusters and clouds that offer various services to users. A discussion of different types of distributed systems is presented next.

1.5.1 Nodes Communicating via Remote Procedure Calls (RPC)

One of the basic types of distributed systems is a distributed memory system with multiple nodes communicating with other nodes interconnected with each other through an intercommunication network. Processes running on the nodes communicate with one another using RPC. An RPC functions like a conventional procedure call except that the calling process and the called procedure exist on two different nodes. The procedure call originates from a client node while the called procedure resides on the server node. One of the goals of an RPC is to make the invocation of a remote procedure look like an invocation of a local procedure to the programmer. Since the called procedure runs on a different node the underlying infrastructure needs to convert these remote producer calls into messages that are exchanged between the client and the server side with the help of their respective operating systems. A schematic diagram showing both the client and server sides is presented in Figure 1.5.

The components on the client side are presented in Figure 1.6.

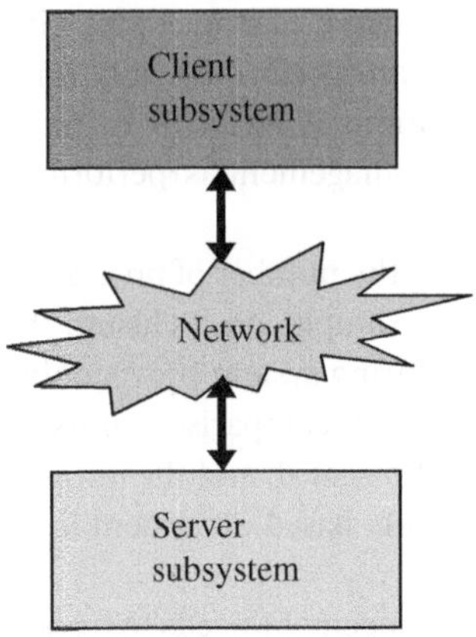

Figure 1.5 Remote procedure call.

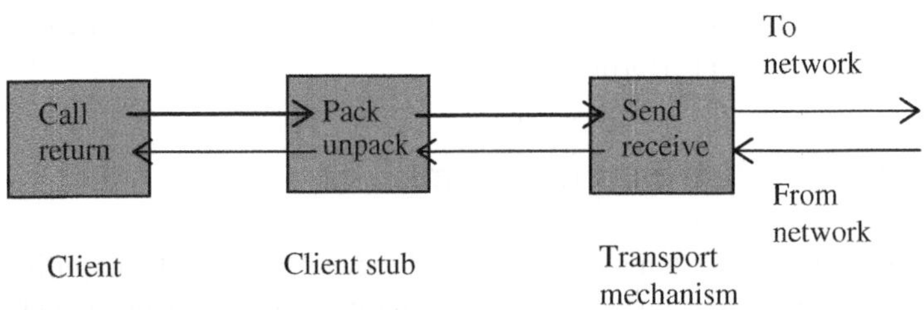

Figure 1.6 RPC architecture (client side).

The various components of the system depicted in the figure are described.

Client Stub: aids in converting the procedure-based communication between the caller module (client) and the called module (server) to message-based communication. It performs two operations:

Marshalling (packing) of data: that corresponds to converting the procedure name and arguments into a message to the server.

Unmarshalling (unpacking) of data: that corresponds to converting the message coming from the server to the results to be returned to the process that called the remote procedure.

Client Transport Mechanism: is the communication network software that sends the marshalled data to the server and receives the message returned by the server transport mechanism.

The various components of the system on the server side shown in Figure 1.7 are described.

Server Transport Mechanism: It is the communication network software that receives the message from the client side and sends the marshalled data from the server stub to the client transport mechanism.

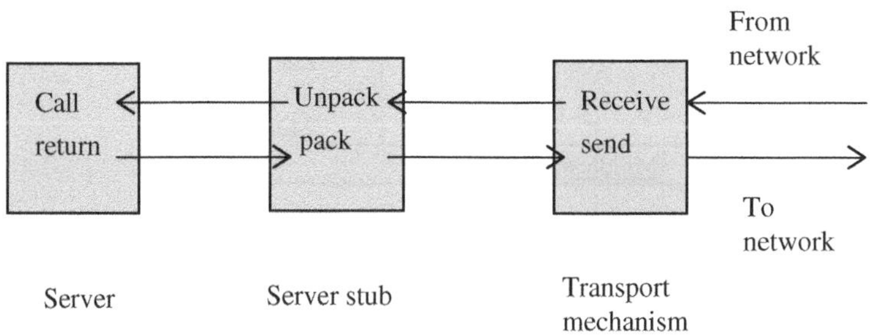

Figure 1.7 RPC architecture (server side).

Server Stub: aids in converting the procedure-based communication between the caller module (client) and the called module (server) to a message-based communication. It performs two operations:

Marshalling (packing) of data: that corresponds to converting the result returned by the remote procedure into a message.

Unmarshalling (unpacking) of data: that corresponds to converting the message coming from the client and converting it to the procedure name and list of arguments passed on to the server process.

Server Process: executes the called remote procedure. The procedure name and list of arguments passed by the client are used to make a procedure call on the server-side machine.

Upon rising the server node registers all the remote procedures it hosts with the name server. At link time the client is bound to a procedure that communicates with the name server at run time to find the server details so that it can communicate with the appropriate machine that hosts the remotely called procedure.

1.5.2 Distributed Object Based Computing Systems

Distributed object computing (DOC) combines the advantages of distributed systems such as concurrency and reliability with those of object-oriented computing such as reuse. Heterogeneity is often an important attribute of such systems. The client, for example, may be written in C++ running over the Linux operating system where as the server that is being called may be written in Java running on top of the Windows operating system. A middleware is often used to provide interoperability in such heterogeneous systems. Different standards for middleware have been available including Common Object Request Broker Architecture (CORBA) (CORBA 2023) and Distributed Component Object Model (DCOM) (DCOM October 2022). The first standard is from the Object Management Group (OMG) while second one is proposed by Microsoft. The role of the middleware in DOC systems

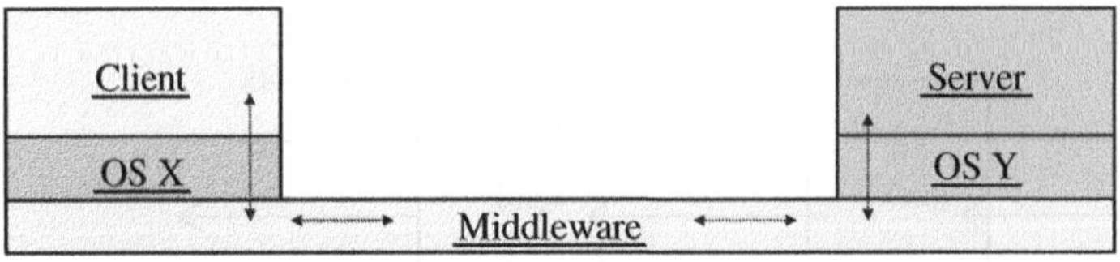

Figure 1.8 Middleware for providing interoperability.

is to provide the necessary translations such that the client can effectively communicate with the server. An example CORBA-based client–server system is shown in Figure 1.8.

The client object (running on top of operating system X) typically invokes a method in the server object (running on top of operating system Y). Since the client and servers may be built using different technology the native data format used in the client and the server may not be the same. The middleware translates the data on the client side into an intermediate form (e.g. common data format [CDF] in CORBA) before transmitting it on the communication network. After receiving this CDF data on the server side, it is translated into the native data form understandable by the server system. Although such translations introduce system overheads the CORBA middleware-based system allows the diverse client and server objects to communicate with one another.

1.5.3 Parallel and Cluster Computing

Another step in the evolution of distributed systems involves parallel and cluster computing systems. An example of a parallel system is a shared memory multiprocessor system that incorporates a number of processing elements connected with shared memory. Processes running on the individual processors can intercommunicate with one another through shared variables placed in the shared memory. Being able to process multiple parts of the problem concurrently this parallel processing system can speed up the data processing operations. Such multiprocessor systems not only can be used to decrease the processing latency of single applications but can also be *multiprogrammed* to run multiple parallel applications simultaneously on the system (Majumdar et al. May 1988). Resource management techniques that include static, dynamic as well as semi-static scheduling are introduced in (Majumdar et al. May 1988). In static scheduling the number of processors allocated to an application is fixed and cannot change during the lifetime of the application. For dynamic scheduling, there is no restriction to the number of processors allocated to an application that can change dynamically at any point in time. Semi-static scheduling is something in between. Changes to the number of processors allocated to a job are not unrestricted as in the case of dynamic scheduling but can be made when specific changes in the system state or in the execution behavior of an application occurs on the system (Majumdar et al. 1991).

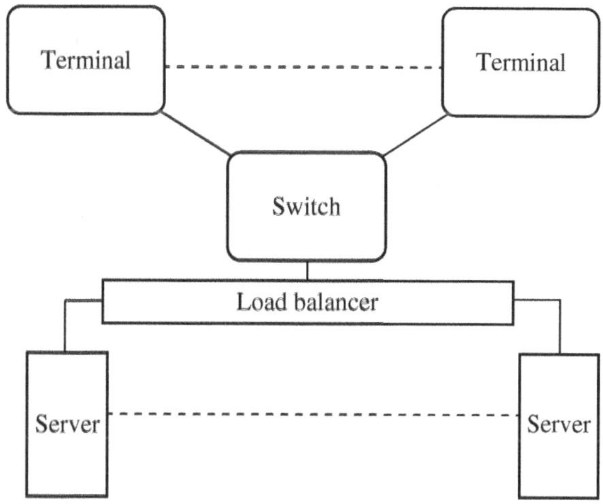

Figure 1.9 Example of cluster computing.

A computing cluster is a collection of multiple independent servers intercommunicating with clients via a local area network or switch for example. Enhanced performance, load balancing, and enhanced reliability are two important attributes of cluster computing. By using multiple nodes in parallel, a cluster can process data much faster in comparison to a single node-based system. If one of the nodes in the cluster fails another node can take over the work that was being performed by the failed node. Clusters can also perform load balancing by distributing the load evenly among the participating nodes so that a performance bottleneck resulting from the existence of a saturated server, while there are lightly loaded nodes in the cluster can be avoided (see Figure 1.9). The terminals in the figure are used by the clients to send work to the cluster. The load balancer is responsible for distributing the work to individual servers. A cluster often employs a master–slave architecture. The master node is responsible for allocating slave nodes to parallel programs, managing a queue of submitted jobs and the user interface. The cluster often includes libraries such as the ones that provide message-based communication facilities. An example of a cluster is the Beowulf cluster (Kumar May 2009).

These are typically a local solution because shared memory multiprocessor systems or clusters are typically housed in a single office in an enterprise or in an institution. Systems that can span multiple geographic locations are discussed in the next section.

1.5.4 Service-Oriented Architecture

Service-oriented architectures (SOAs) concern a distributed system on which applications take the aid of well-defined services for achieving the application functionality. A service abstracts a well-defined repeatable activity and SOA refers to a set of services that interact with one another via a given communication

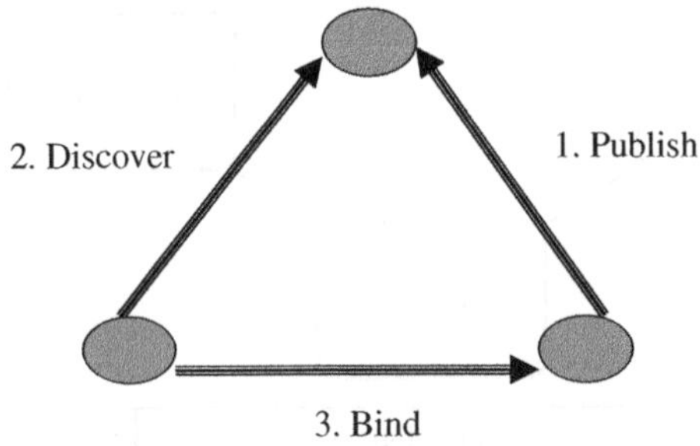

Figure 1.10 Use of a service in an SOA-based system.

protocol. The interactions result in data exchange between two services. Multiple services in an SOA-based system cooperate with each other to achieve a given objective. Before using a service, the service requester needs to discover a service with the help of a registry (see Figure 1.10). Upon rising, a server registers its services with a third-party service registry that is also referred to as a service broker. The main function of the registry is to make information regarding a service including its unique location available to a service requester. After discovering a service the service requester binds with the service provider to use one of its services.

Key characteristics of services include the following:

- They provide a uniform abstraction of various entities such as processes, and databases.
- Their interactions are message-based: messages are exchanged between the service provider and the service requester.
- While providing a given service the service provider may invoke additional services.
- Service implementation is abstracted and not visible to the requester.
- Service description: a service has a machine-processable (metadata) description available in the registry.
- The service provider and service consumer may be deployed on their own platforms. They interact with the help of a platform-neutral message format.

Web services (W3C February 2004) can be used to build SOA applications. A Web service supports machine-to-machine interactions over a network. It provides an API specified by using the Web Service Description Language (WSDL) (W3C February 2004). Web services can be discovered using the Universal Description, Discovery, and Integration (UDDI) protocol. They often employ the Simple Object Access Protocol (SOAP) (W3C April 2007) for exchanging messages using typically the platform neutral Extended Markup Language (XML) (W3C 23 2023) data format.

1.5.5 Grids

Clusters are often homogeneous and the nodes in the cluster are mostly of the same type. A computational *grid*, on the other hand, is an infrastructure that allows a heterogeneous set of machines provided by a group of collaborating people/institutions to form a virtual organization (VO). The machines can be accessed by any authenticated user of this VO. There is no restriction on the type of machines that form this VO and thus different members can provide access to different types of machines.

A grid is a distributed system that provides coordinated large-scale heterogeneous resource sharing and problem solving in dynamic, multi-institutional VOs.

It is used for distributed computing and massive data sharing/processing, as a common platform and way for utility and service computing. Resources may include computers, databases, storage devices, light paths, sensors, and measurement equipment.

A grid is described as a system that is able to coordinate "resources that are not subject to centralized control" use "standard, open, general-purpose protocols and interfaces" and "deliver nontrivial qualities of service." (Foster July 2002).

A layered architecture for the grid is available in (Foster et al. 2001). A short discussion of the roles of the different layers shown in Figure 1.11 is presented.

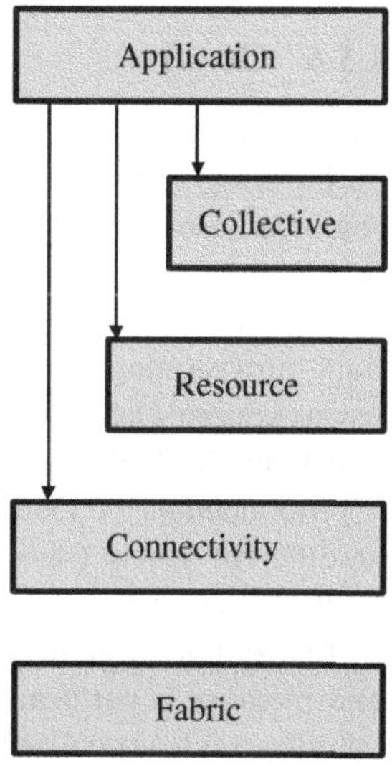

Figure 1.11 Layered architecture of the grid.

Application Layer: provides an abstraction of a VO. It uses the collective, resource, connectivity, and fabric layers that are described next.

Collective Layer: provides access to and coordinates interactions across a collection of resources. It is also responsible for the coordination of multiple resources. It includes services for resource discovery, allocation, co-reservation of multiple resources, brokering, and process scheduling.

Resource Layer: provides single resource sharing and focuses on sharing of individual resources: negotiating access and controlling use of the resource. It performs access control and relies on the authentication provided by the connectivity layer. It builds on the connectivity layer protocols to provide initiation, monitoring, accounting, and billing for single resources.

Connectivity Layer: provides core communication and authentication protocols. It concerns communication requirements between fabric layer resources including transport, naming, routing, security, etc. Its services include providing communication protocols for supporting grid transactions that require the use of multiple resources as well as for communication with and access of a given resource from a remote location.

Fabric Layer: provides the resource-specific services. A resource may be a computer, network or a logical entity such as a distributed file system. This layer concerns the controlling of resources locally at a specific site. One of its objectives is to facilitate the sharing of resources within a VO. This includes facilities for querying the state of a given resource.

1.5.6 Clouds

Cloud computing often refers to the delivery of computing as a service. It provides shared (virtualized) resources, software, and information as a utility over a network (typically the Internet). A cloud is characterized by a number of features: virtualization; provisioning resources on demand; elasticity; pay-as-you-go; QoS/service level agreement (SLA); and scalability. Each of these is briefly discussed.

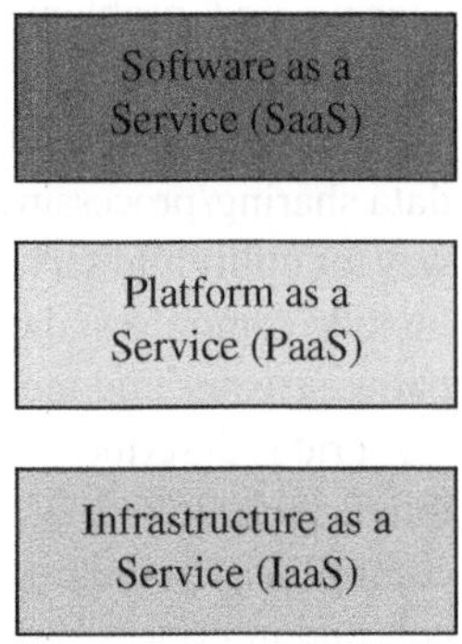

Figure 1.12 Three levels of virtualization.

Virtualization: A cloud can provide virtualization at different levels (see Figure 1.12). Each of these levels is associated with a specific class of service. Virtualization of hardware resources such as CPU and memory is performed for use by multiple users. Offering such virtualized resources is referred to as *Infrastructure as a Service (IaaS).*

Including facilities such as those for program design and development that include an IDE, for example, application testing and deployment leads to a service bundle that is referred to as *Platform as a Service (PaaS).* The level above PaaS concerns providing specific application functionalities such as customer relationship management (CRM) and is referred to as *Software as a Service (SaaS).*

Resources on Demand: The resources can be acquired on demand. That is, the user can acquire one or more resources as and when needed. These are to be used by the user for the required time period and then returned to the cloud service provider. The requested number of resources are provisioned by the cloud service provider at the requested time.

Elasticity: The resource pool provided by a cloud is said to be elastic. The number of resources in the pool can be increased or decreased dynamically based on the user's requirements.

Pay as You Go: This price model used for clouds allows the user to pay only for the duration of time they have used the resources from the cloud. If the user uses the services from the cloud in the future the user is billed again based on the resource usage during that period of use.

Quality of Service (QoS)/Service Level Agreement (SLA): In certain systems, it is possible to associate QoS requirements with resource usage. Once the user request is accepted, the cloud service provider is said to have been associated with an SLA with the user.

Scalability: Since a cloud data center is typically associated with a very large number of resources the environment offers scalability for user applications.

Clouds provide several advantages to its clients and are described next:

- Providing resources on demand using a pay-as-you-go model can reduce the user costs related to IT. The virtualized resources are "rented" by the clients temporarily and they need to pay only for the duration of time they have used the resources. This is particularly important for startup companies who do not need to lock their funds in getting infrastructure but "rent" the required number of resources when needed. Since computing infrastructure is provided by the cloud provider, it implies a reduction in in-house equipment maintenance. Not having to maintain computing equipment can lower the IT management cost for all companies and institutions.
- Enhances system performance by providing access to multiple resources. By deploying multiple resources a cloud can lead to the speeding up of a parallel application and/or increasing system throughput by executing more jobs per unit time.
- Getting resources on demand leads to an elasticity in the resource pool. This allows the dynamic adjustment of the number of resources to temporarily handle spikes in the workload.
- Potential reduction in energy consumption: Since the computer infrastructure is maintained by the cloud service provider the in-house energy cost is lowered for the client enterprise. By housing hundreds/thousands of computers in one enclosure cloud data centers consolidate air-conditioning requirements, thus amortizing the cooling cost over multiple equipment and are thus said to be in-line with the concepts of green computing.

The advantages provided by the cloud are accompanied by a number of challenges. These are briefly described.

Achieving Adequate Security and Privacy: Since the same infrastructure is used by multiple persons and competing enterprises it is crucial to ensure that privacy and security are maintained and the data of one user is not accessible to another user.

Providing Effective Resource Management Techniques: Clouds provide users with access to a large resource pool. Appropriate management of these resources is crucial for effectively utilizing the power of this underlying resource pool. User requests are often associated with an SLA that includes the execution time of jobs, their earliest start time, and a deadline. User estimates of job execution times are often error-prone. Resource management is further complicated by the requirement of handling with such inaccuracy in user estimates of job execution times. A cloud data center typically comprises hundreds/thousands of servers each of which is running its own operating system. Moreover, the resource pool is dynamic and new resources may be added to and existing resources may be removed from the pool. Allocating resources to jobs in such a heterogeneous environment is a challenging problem.

Adequate Facilities for Monitoring/Management: A user application running on a cloud may comprise multiple components each of which may be running on a different machine. Monitoring the progress of the application requires the monitoring of its different components that adds to the complexity of the operation.

Interoperability Among Multiple Clouds: It is desirable for a user to be able to run an application on one of many systems provided by multiple service providers. This avoids the vendor lock in situation in which a user is forced to keep on using the same cloud for running the application even though better price/performance may have become available from a different cloud service provider.

1.5.7 Edge Computing

Edge computing (EC) refers to a form of distributed computing that provides both data storage and computing facilities close to the user or sources of data (edge devices). This concept is useful when computing is performed in a hierarchical fashion where some of the computing is performed on the edge of the network by edge nodes and some at a backend facility such as a cloud data center (see Figure 1.13). The edge devices that generate the computation requests are directly connected to the edge nodes that in turn are connected to the backend cloud. This hierarchical architecture is used to improve responsiveness to the user by avoiding the network delays that will occur if all the data needs to be transferred over the network from the edge devices to a backend cloud for processing. In many EC systems, the resources at the edge are less powerful in comparison to the ones at the backend cloud. Typically, time-sensitive processing is performed at the edge to avoid network delays, while computations without short response time requirements are delegated to the backend cloud.

By performing computations and data storage at multiple levels EC makes use of a larger number of resources that can lead to an improvement in the overall speed of the system. Moreover, when computations are performed by the edge nodes that are close to the edge devices, the volume of data transfer over the network can be reduced significantly, thus lowering the use of network bandwidth and decreasing inter-process communication delays. Reducing the computing load on

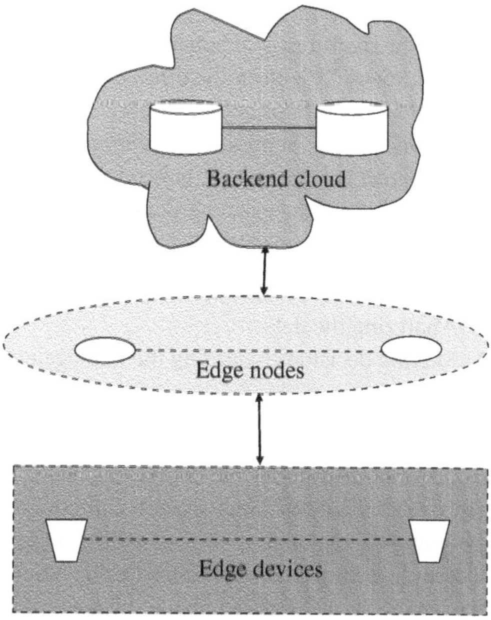

Figure 1.13 Edge computing.

the backend cloud by performing some computations at the edge can also lead to an improvement in the scalability of the system. Keeping the data local tends to improve data privacy. Enhanced security mechanisms such as special encryption may be required for transferring the data over the network. EC systems often use resource-constrained devices such as smartphones at the edge. Their limited computing power, memory capacity, and energy need to be considered carefully while building such an EC system.

1.5.8 Smart Facilities

The advancement in distributed systems in general and in cloud computing and IoT-based techniques in particular are fueling the growth of smart systems. Examples of such systems include smart infrastructure or cyber–physical systems, smart transportation, smart commerce, smart health care, smart machinery, and smart entertainment. Development of such smart facilities is paving the way for achieving the vision of smart cities in the world. Bridges are getting equipped with sensors that produce the data for keeping track of their structural health. Sensors are being used in buildings to collect data on their occupancy level and ambient temperature and light intensity in rooms for efficient operation of such buildings. Smart commerce refers to e-commerce systems including mobile commerce that puts the e-commerce capabilities directly in the hands of users by leveraging wireless technology based mobile systems. Smart health care systems enable the monitoring of patients remotely whereas smart entertainment systems such as smart TV allows the direct access of various sources of movies and news items by the user. Smart machinery such as industrial robots that can operate autonomously or semi-autonomously helps in achieving computer integrated manufacturing leading to enhanced efficiency and productivity.

Irrespective of the exact nature of the smart facility the three following steps of operation are often used.

1. Collecting data that reflects the current state/health of the facility.
2. Analyzing the data.
3. Making decisions regarding the management/maintenance of the facility.

Depending on the specific system sensors, data analysis tools and databases containing archival data about the previous maintenance/operation history of the respective facility as well as computing platforms/servers are required for carrying out these operations. Such system components are often scattered across a large geographical region that may span different locations. An effective distributed system architecture and techniques for the handling of such geographical dispersed resources is crucial for achieving the goals of smart facilities management.

1.6 Summary

The growth in both processing and communication technology in the past decades has fueled the growth of distributed systems that provide both a higher performance as well enhanced reliability in comparison to a single-processor-based system. In this introductory chapter, we have discussed earlier distributed systems that include a small number of processes sharing one or more processing elements. Techniques for process communication by using RPC and for object interaction in middleware-based heterogeneous interoperable systems were described. Interoperability provided by incorporating the web services technology in service-oriented architecture-based systems were discussed. The chapter also includes an introduction to parallel and cluster-based systems, as well as grid, cloud, and EC. The chapter also includes an introduction to smart facilities that incorporate processing, communication as well a sensor technology.

The book concerns resource management. A few general principles of resource management on distributed systems that will be used in the subsequent chapters were introduced in this chapter. The rest of the book is divided into two parts. The first part concerns resource management in distributed computing systems, whereas the second concentrates on resource management in data-intensive distributed systems. The focus of each chapter contained in both parts of the book is described in the next subsection that aims to map the key concepts to the chapters and describe how they are related to one another.

1.6.1 Book Components

This book concerns resource management in distributed systems that can be classified into two types: computing-intensive systems and data-intensive systems. Each system type is the focus of attention in a given part of the book. Part 1 (Chapters 2–5) focuses on issues underlying distributed computing-intensive systems whereas Part 2 (Chapters 6–10) is concerned with distributed data-intensive systems.

Part 1: Chapter 2 focuses on concurrency in the execution of processes and threads that are the fundamental units of computation performed on a distributed system. The attributes of the computation that have an intimate relationship with system behavior and performance are described. Important metrics that capture the impact of these characteristics on performance are introduced. The characteristics and performance metrics described in this chapter can often be used for explaining the behavior of the application when it is executed as well as in the management of system resources on which the application may be deployed.

There are two important components of resource management: allocation and scheduling that are the subject of attention in Chapter 3. Allocation refers to the mapping of processes/threads to processing resources. An allocation algorithm determines which process in an application will be executing on which of the system resources. *Scheduling* concerns the order in which the various processes/threads running on the system are executed. Various algorithms for both allocation and scheduling that include both optimal and heuristic techniques are described. Special algorithms that try to ensure the meeting of application deadlines are included. Resource management on grids and clouds as well as on client–server systems is discussed.

The resource management techniques discussed in the previous chapter often assume that exact *a priori* knowledge about applications; workloads and system resources are available. Real systems, however, are often subjected to uncertainties associated with applications, workloads, and resource characteristics. Estimates of task execution times required by many resource management algorithms are often prone to error. Appropriate error handling techniques are required for incorporating robustness into the resource management algorithms. Large distributed systems such as cloud data centers can comprise hundreds or thousands of resources. The local operating systems used on the resources can be different from one another. The resource management for jobs often depend on the allocation and scheduling algorithm at the system level as well as the scheduling algorithm used by the local operating system running on the resource.

Handling such variability in local resource management strategies used by the different local operating systems on resources complicates resource management even further. Handling both of these types of uncertainties is the subject of attention in Chapter 4 that describes techniques for handling inaccuracies associated with task execution times and for making effective resource management decisions without a detailed *a priori* knowledge of local resource management policies used at the various resources on a cloud.

The discussion of resource management techniques presented so far is based on the assumption that there is a fixed number of resources on the system. Such an approach can lead to an ineffective system that is often subjected to a variability in workload intensity. The system with a fixed number of resources can therefore give rise to an under provisioned system exhibiting poor performance when the system load is high and an over provisioned system with idle resources when the workload is low. Auto-scaling is a technique that dynamically controls the number of system resources based on the current system load and is the subject of attention in Chapter 5. Three types of auto-scaling techniques are discussed: reactive auto-scaling that changes the number of system resources in reaction to the change in the system workload, proactive auto-scaling that predicts the future system workload and decides on the required changes in the number of resources in advance as well as hybrid auto-scaling that combines reactive and proactive auto-scaling.

Part 2: Chapters 6–10 focus on data-intensive systems that include systems processing stored data as well as systems that handle streaming data. Systems processing stored data include data analytics platforms used for the processing of big data. Resource management techniques for the well-known MapReduce platform are described in Chapter 6. The system is subjected to a workload characterized by MapReduce jobs associated with SLAs, such as job completion deadlines. Errors are often associated with task execution time estimates that are provided by the user. Discussion of such inaccuracies in the data provided by the user and resource management techniques for handling the errors are discussed.

Energy consumption of data centers processing MapReduce jobs is an important concern in the context of resource management. Reducing the energy consumption of the system resources processing MapReduce jobs is important for controlling the operating expense for the data centers. Energy-aware resource management techniques that not only meet the user-provided SLAs but also consider the controlling of energy usage are described in Chapter 7.

Analysis of streaming data is discussed in Chapter 8. Streaming data is characterized by both the volume and velocity of the streaming workload. One of the challenges of streaming data is the dependency of workload processing and resource management on data content. Two types of streaming systems are considered in this chapter. The first concerns priority-based message processing, and the second considers complex event processing. Priority-based resource management is required in systems where certain "high-priority" data items detected in the incoming stream need to be processed in advance of the others. Complex event processing is used for analyzing multiple streams of data in real time to generate alerts or to indicate possible opportunities to be exploited on the system. Resource management for systems processing streaming data is discussed in this chapter. In addition to resource management techniques platforms for processing streaming data such as Apache Spark and Apache Storm are also discussed.

Data is an important resource, the processing of which can lead to substantial revenue for enterprises and managing of such data resources is the subject of attention in Chapter 9. Searching is an important operation performed during the processing of big data. Large volumes of such data can increase the search time. Two approaches for addressing this problem are discussed in Chapter 9: (i) data indexing and (ii) data filtering. Data indexing is a technique for reducing the latency of performing searches on a document set with a large volume. Different indexing techniques and their impact on system performance achieved during data searching are described. Another method of speeding up data processing is through the reduction of data volume by performing data filtering. Such an operation reduces the data size by storing only those data items that are of interest to the user. Data filtering techniques that are often parallelized to reduce the filtering overhead are discussed both in the context of stored data as well as streaming data in this chapter.

Special resource management techniques are required for processing streaming data in the context of sensor-based subsystems that are often incorporated into systems that use IoT technology, such as cyber–physical systems. Resource management techniques for senor networks are presented in Chapter 10 that provides a description of a number of scheduling and allocation policies and their performance.

Chapter 11 summarizes the key topics presented in the different chapters of the book.

References

CORBA, 2023, www.CORBA.org

DCOM, "Microsoft, MS-DCOM: Distributed Component Object Model (DCOM) Remote Protocol," October 2022, https://learn.microsoft.com/en-us/openspecs/windows_protocols/ms-dcom/4a893f3d-bd29-48cd-9f43-d9777a4415b0

I. Foster, "*What is the Grid? A Three Point Checklist*", Argonne National Laboratory & University of Chicago, Chicago, USA, July 2002.

I. Foster, C. Kesselman, S. Tuecke, "The Anatomy of the Grid: EnablingScalable Virtual Organizations," *International Journal of High Performance Computing Applications*, Vol. 15, Issue 3, 2001, pp. 200-222.

L. Kleinrock, "*Queueing Systems, Volume I*", Wiley-Interscience, NJ, USA, 1975.

S. Kumar, "*Building a Beowulf Cluster in Just 13 Steps*", May 2009, https://www.linux.com/training-tutorials/building-beowulf-cluster-just-13-steps/

S. Majumdar, D.L. Eager, R.B. Bunt, "Characterisation of Programs for Scheduling in Multiprogrammed Parallel Systems", *Performance Evaluation*, Vol. 13, Issue 2, October 1991.

S. Majumdar, D.L. Eager, R.B. Bunt, "Scheduling in Multiprogrammed Parallel Systems", *Proc. ACM SIGMETRICS Conference on Measurement and Modeling of Computer Systems*, Santa Fe, USA, May 1988.

A. Silberschatz, P.B. Galvin, G. Gagne, "*Operating Systems Concepts*", Wiley, NJ, USA, 2014.

W3C, "*W3C Recommendation (Second Edition)*", April 2007, https://www.w3.org/TR/soap/

W3C, "Web Services Glossary", February 2004, https://www.w3.org/TR/2004/NOTE-ws-gloss-20040211/#webservice

W3C 23, "Extensible Markup Language (XML)", 2023, https://www.w3.org/XML/

2

Characterization of Parallelism in Applications

2.1 Introduction

Both the software and hardware components of parallel and distributed systems are responsible for the overall system behavior, performance as well as energy consumption. The hardware architecture as well as the characteristics of the CPUs are examples of hardware characteristics of the system. The software components of the system include both application and system software. Examples of system software are the operating system, compiler, and assembler. The operating system manages the hardware and software components, facilitates the users' interaction with the system, and manages the various system resources. The compiler and assemblers are responsible, respectively, for translating an application program written in a high-level language or assembly language into object code. Application programs, on the other hand, achieve the desired functionality that the user wants to achieve. Parallelism in both hardware and software is used for producing high system performance and/or achieve the distributed and/or concurrent execution required by the application. This chapter focuses on the characteristics of parallel/concurrent software.

Characteristics of parallel applications have a strong relationship with their execution behavior and system performance achieved when they execute on the system. Characterization of parallelism in an application refers to the identification of those attributes (characteristics) of the application that have an intimate relationship with system behavior and performance. Such characteristics can often be used for explaining the behavior of the application when it is executed as well as in the management of system resources on which the application may be deployed. Both graph-based characterizations as well as single-point attributes of parallel applications are available in the literature. Before discussing these

Resource Management on Distributed Systems: Principles and Techniques,
First Edition. Shikharesh Majumdar.

characteristics, a popular model called the *precedence graph model* for describing parallel applications will be introduced. The model will be useful in the discussion of the parallelism characteristics provided in this chapter. This will be followed by a discussion of both graph-based and single-point characteristics of parallel applications.

Parallelizing applications can often give rise to an increase in energy consumption. Consider, for example, the use of multiple processes that run on multiple processors. The increase in the number of processors needed can increase energy consumption. The increase in energy consumption is not only due to the energy consumed by the processing elements but also by the energy consumption of the air-conditioning units deployed to cool the space housing the equipment. To address the relationship between parallel software and energy consumption, a number of energy characteristics are also discussed. The parallelism characteristics are discussed first, and this discussion is followed by a discussion of the energy characteristics of the application.

This chapter, which focuses on the parallelism characteristics of applications, is divided into a number of sections. Each of these sections concerns one of the learning outcomes of the chapter. The precedence graph is a well-known model for concurrent computations that run on computing resources as described in Section 2.2. The parallelism characteristics of applications are described next. Two types of characteristics are described. Graph-based characteristics are discussed in Section 2.3, each subsection of which describes a specific characteristic. Single-point characteristics are described in Section 2.4. Section 2.5 describes the various performance metrics that are important in the context of parallel and distributed systems, whereas Section 2.6 discusses the relationship between parallelism characteristics and performance. The characteristics and the metrics related to the energy used by the resources that are important in the context of resource management are discussed in Section 2.7. Section 2.8 concludes the chapter and provides a summary of the material presented in the chapter.

2.2 The Precedence Graph Model

The precedence graph model is a very popular and widely used model for characterizing precedence relationships among the constituent tasks in a parallel/concurrent application. A precedence graph is an acyclic graph that comprises a set of vertices $\Pi = \{t_1, t_2, \dots t_j, \dots, t_m\}$ in which vertex t_j models a task/process in the application that comprises m tasks. The precedence relationship between two tasks t_j and t_k describes whether they need to be executed in a sequence. If t_j must be completed before t_k then a precedence relationship is said to exist between

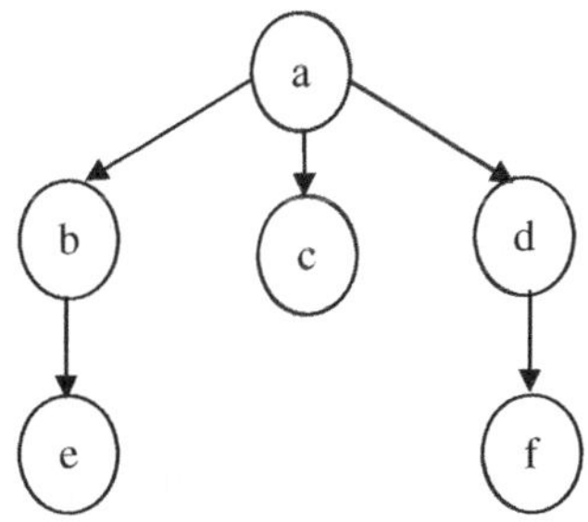

Figure 2.1 Example precedence graph.

these tasks. This precedence relationship is captured in a directed edge from t_j to t_k. The set of precedence relationships for the entire application is thus captured in the partial order:

$$P = \{(t_j, t_k)\} \text{ where } j = 1 \ldots m$$

Each member of the set on the right-hand side of the previous equation depicts a precedence relationship between a specific pair of tasks. If no such precedence relationship exists between a given pair of tasks it means that these tasks can execute concurrently. In addition to the textual notation introduced, a precedence graph is often presented in the form of a diagram. The diagram of an example precedence graph characterized by:

$\Pi = \{a, b, c, d, e, f\}$; $P = \{(a,b), (a,c), (a,d), (b,e), (d,f)\}$ is presented in Figure 2.1.

In Figure 2.1, tasks b, c, and d are successors of task a implying that the execution of task a must complete before tasks b, c, and d can start execution. Task a is also referred to as the predecessor of tasks b, c, and d while tasks b, c, and d are successors of task a. Similarly, b is the predecessor of task e and task d is the predecessor of task f. Since there is no precedence relationship among tasks b, c, and d they are said to be concurrent. Similarly, tasks c, e, and f are concurrent and are allowed to execute at the same time. Task a does not have any predecessor and can execute as soon as the application starts running on the system.

2.3 Graph-Based Characteristics

Three popular graph-based characterizations of parallel applications (Sevcik 1989) are presented. The first, called parallelism, profile can be computed when the application is executed on an *unbounded number of processors*. This means that there is no dearth of processing resources and if any task becomes ready to run then there is an available processor on which the task will be able to run.

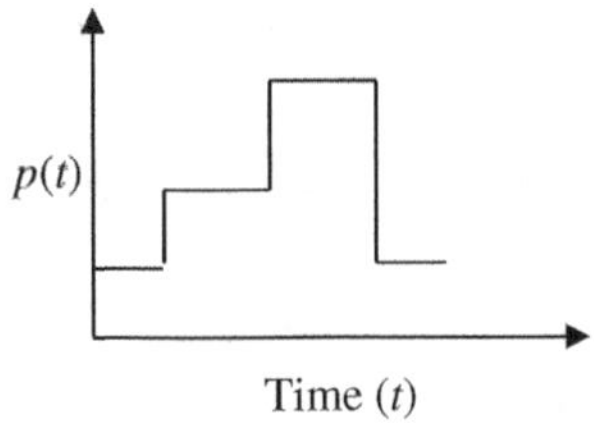

Figure 2.2 Parallelism profile.

2.3.1 Parallelism Profile

The parallelism profile of an application captures the degree of parallelism in the application when it executes on a system comprising an unbounded number of processors. Parallelism profile is presented as a graph with time on the X-axis and the number of tasks/processes, $p(t)$ that is concurrently active at the given point in time, t, on the Y-axis (see Figure 2.2).

An example parallelism profile for the application presented in Figure 2.1 is presented in Figure 2.3. The execution times for the tasks used in this example are presented in Table 2.1.

Note that with an unbounded number of processors, each task is able to run as soon as all its predecessors are complete, leading to the $p(t)$ versus t graph presented in Figure 2.3. When the application starts running ($t = 0$), only one task is running on the system leading to a degree of parallelism ($p(t)$) of 1. When task a finishes ($t = 1$) b, c and d start running on the system. The degree of parallelism thus increases to 3 and continues at this level until $t = 3$ when b and c are complete. The degree of parallelism drops to 2 at $t = 3$ and remains at this level until $t = 4$ when task e completes. Task d continues to run for another 1 unit of time

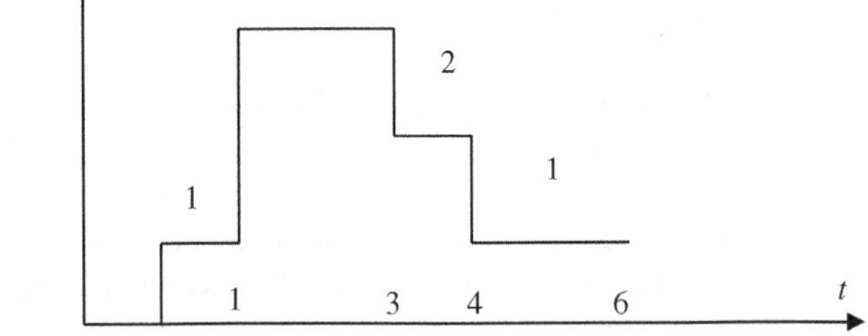

Figure 2.3 Parallelism profile for example application of Figure 2.1.

and task f is started after d completes at $t = 5$. The application finishes with the completion of task e at $t = 6$. Thus, a degree of parallelism of 1 is maintained from $t = 4$ to $t = 6$.

2.3.2 Shape

The parallelism profile provides a detailed description as it captures the degree of parallelism for the entire run time of the application. A shorter summary of the degree of parallelism in an application is provided by the shape function. This characterization is based on the proportion of time each degree of parallelism is exhibited during application execution. Let, p_i be the proportion of time during which the degree of parallelism is exactly equal to i. Note that p_i can be computed from the parallelism profile as:

$$\left[\sum_t p(t) = i\right] / T(\infty),$$

where the numerator is the total duration of time during which $p(t) = i$ and $T(\infty)$ is the completion time of the application when the number of processors is unbounded.

The shape is then defined as a set $\{P_i\}$ where:

$$P_i = \sum_{j=1}^{i} p_j.$$

The shape an application can be presented in a graphical form, as shown in Figure 2.4. Note that the shape for an application always ends with $P_i = 1$.

As an example, the shape of the application is characterized by the precedence graph in Figure 2.1 and the task execution times presented in Table 2.1 is presented next. The duration of time for which a given number of tasks (i) is executed on the system when an unbounded number of processors is available is presented in Table 2.2. The values of p_i and P_i are then computed using this table (see Table 2.3). The resulting shape for the application is given in Figure 2.5.

Figure 2.4 The shape for an application.

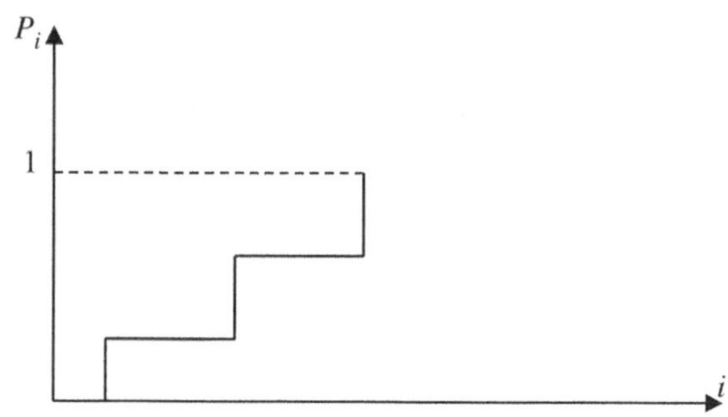

Table 2.1 Task execution times for application from Figure 2.1.

Task	Execution time
A	1
B	2
C	2
D	4
E	1
F	1

Table 2.2 Execution durations for tasks.

Tasks	Execution duration	No. of tasks
A	1	1
b, c, d	2	3
d, e	1	2
D	1	1
F	1	1

Table 2.3 Computation of probabilities for generating the shape graph.

Degree of parallelism (i)	p_i	P_i
1	3/6	3/6
2	1/6	4/6
3	2/6	1

2.4 Single-Point Characteristics

Graph-based characteristics for parallel applications provide detailed information about the application. However, they also take space for storage and time for processing when used in resource management.

Single-point measures that summarize the parallelism characteristics are thus also used for the characterization of parallelism. These characteristics are often apt

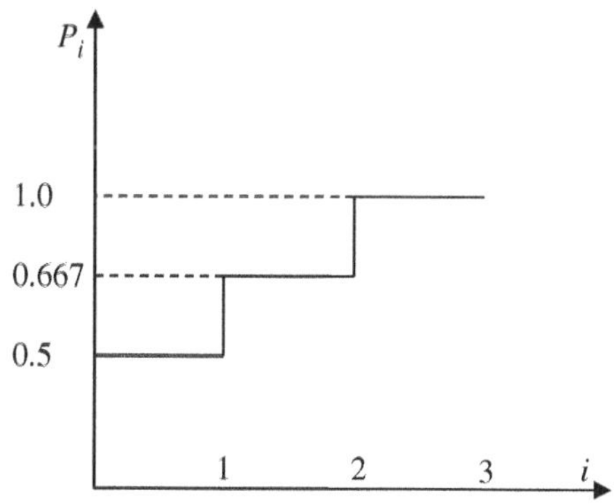

Figure 2.5 Shape for example application.

for use in the provisioning and management of resources on which the application executes. These characteristics not only capture the degree of parallelism in the application but also can be used to provide a prediction of performance that is going to be achieved when the respective software executes on the system. The different single-point characteristics that are discussed in the literature (Sevcik 1989) are introduced first. Their impact on system performance is discussed in the following sections.

2.4.1 Maximum Parallelism

The maximum parallelism (m) for an application is the maximum number of processes that can execute simultaneously when the application is executed on a system with an unbounded number of processors. m can be computed from the parallelism profile discussed earlier as the maximum value of $p(t)$ observed during the execution of the application when the number of processors is unbounded.

2.4.2 Fraction of Sequential Work

Each application typically has some tasks that must be executed sequentially and no other task is allowed to execute on the system while such a task is in execution. Examples include a critical section in a concurrent program when execution of only a single process/task at a given point in time is required for the correct operation of the application.

The total work performed during the application execution, W, is the sum of the execution times for all tasks in the application. Fraction of sequential work (f) is the ratio of the total time T_s during which execution must remain sequential and W:

$$f = T_s/W.$$

Note that sequential execution may be spread over multiple intervals of time during the entire execution of the application and T_s is the sum of the duration for all these intervals of time. f can be computed from the parallelism profile for the application described earlier:

$$f = \text{Area}_1/\text{Area}_s$$

The numerator is the area under the parallelism profile graph for durations when $p(t) = 1$, and the denominator is the total area under the graph. This is an important application attribute because it can be used to predict the highest speedup achievable on a given system.

For the example described by Figure 2.1 and Table 2.1, the value of f is 3/11. This is because, given that the number of processors is unbounded, the parallelism profile (Figure 2.3) shows that the degree of parallelism is only 1 during the period of time 0 to 1 and the period of time 4 to 6. The sum of the execution times for all the tasks is 11.

2.4.3 Average Parallelism

This measure is often used to understand "how parallel" a given application is. *Average parallelism (A)* (Eager et al. 1989) is defined as the average number of processors that are busy when the application is executed on a system with an unbounded number of processors.

Average parallelism is related to the degrees of parallelism exhibited during this execution:

$$A = \sum_i i\, p_{i,} \tag{2.1}$$

where p_i is the proportion of time the degree of parallelism is exactly equal to i and can be computed from the parallelism profile for the application. A for example application characterized by the precedence graph in Figure 2.1 and the task execution times presented in Table 2.1 is computed as:

$$A = 1 * 3/6 + 2 * 1/6 + 3 * 2/6 = 1.833.$$

Note that the p_i values used in this computation are the ones available from Table 2.3.

2.5 Performance Metrics

Several measures for the performance of parallel applications are discussed in the literature. A number of such key performance metrics is discussed next.

2.5.1 Completion Time

The completion time $T(N)$ for a parallel application is the total time it takes to complete on N processors given that a work-conserving scheduling discipline is used. For such a scheduling discipline a processor is never kept idle if there is at least one ready task on the system. $T(N)$ is dependent on the task scheduling discipline used. $T(2)$ for the application characterized by the precedence graph presented in Figure 2.1 with task execution times specified in Table 2.1 running under a First Come First Served scheduling discipline is 8. Each task after becoming ready to execute joins a First In First Out (FIFO) queue. Ties are broken using a lexicographic ordering for task labels. Whenever a processor becomes free the task at the head of the queue is selected to run. The computation of $T(2)$ is demonstrated with the help of the Gantt chart presented in Figure 2.6.

The execution of each task is displayed on the line representing the processor on which it is executed. Note that a different value of $T(N)$ will be achieved if task d was scheduled at time 1 instead of task b.

Two values of completion time, $T(\infty)$ and $T(1)$ have a special significance. The first is the time of completion achieved on a system with an unbounded number of processors whereas the second is the sequential execution time for the application achieved when the application is run on a single processor. Note that, $T(\infty)$ is the shortest completion time for the application. It is the length of the *critical path* (the longest path from the root to a leaf node) for the precedence graph characterizing the application.

2.5.1.1 Processor Sharing Scheduling

Completion time achieved with a Processor Sharing (PS) scheduling discipline that is a special theoretical scheduling discipline used in many analyses (see Eager et al. 1989 for example) is described here. For such a discipline, the processors are *equally shared* among all the ready processes in the system.

The completion time $T(N)$ achieved with a PS discipline can be computed in terms of $T(\infty)$ (Eager et al. 1989):

$$T(N)_{\mathrm{PS}} = T(\infty) \left[\sum_{i=1}^{N} p_i + \sum_{i>N} p_i \right], \tag{2.2}$$

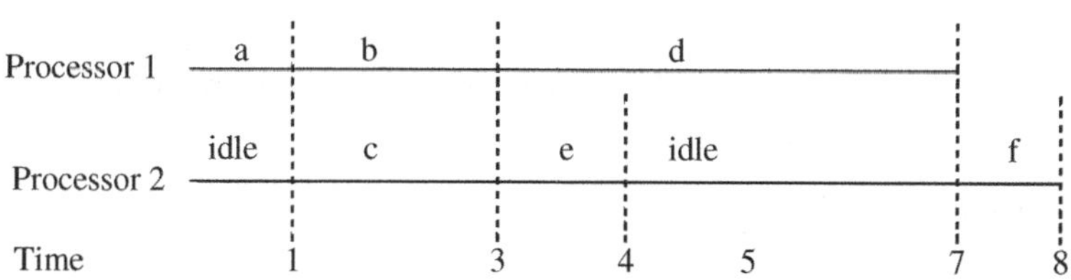

Figure 2.6 Gantt chart for example application for $N = 2$.

where p_i is the proportion of time for the degree of parallelism i observed on a system on which the number of processors is unbounded.

2.5.2 Speedup

The performance improvement resulting from the parallel execution of an application on multiple processors is captured in the speedup $S(N)$ achieved by executing the application on N processors. It is defined as the ratio between the completion time achieved with its sequential execution and the completion time achieved on an N processor system:

$$S(N) = T(1)/T(N)\,. \tag{2.3}$$

Since $T(N)$ depends on the way tasks are scheduled on the N processors, $S(N)$ is also dependent on the resource management policy used. For the 2 processor example corresponding to the execution of the application (characterized by the precedence graph of Figure 2.1 and task execution times described in Table 2.1) captured in the Gantt chart Figure 2.6:

$$S(2) = 11/8 = 1.375.$$

2.5.3 Efficiency

Efficiency $E(N)$ achieved on a system with N processors captures the average processor utilization and is computed as the ratio between the total work W (or $T(1)$) and the total processor time used during the execution of the application on N processors:

$$E(N) = W/(N * T(N)) = T(1)/(N * T(N))\,.$$

Note that since N processors are held for $T(N)$ units of time the total processor seconds that accrue from its execution on N processors is the product of N and $T(N)$. For the 2 processor example corresponding to the execution of the application (characterized by the precedence graph of Figure 2.1 and task execution times described in Table 2.1) captured in Figure 2.6:

$$E(2) = 11/(2 * 8) = 0.6875.$$

Note that $E(N)$ is related to $S(N)$. Rearranging the terms on the right-hand side of the equation for $E(N)$ presented earlier yields:

$$E(N) = \left[T(1)/T(N)\right]/N.$$

Noting that $T(1)/T(N)$ is $S(N)$ we get:

$$E(N) = S(N)/N.$$

2.5.4 Completion Time-Efficiency Profile

Most applications demonstrate a trade-off between speedup and efficiency. Typically, increasing N tends to increase $S(N)$, but this increase in speedup is often accompanied by a decrease in efficiency. This graph-based characteristic of a parallel application is captured in the completion time-efficiency profile, which is a plot of $E(N)$ versus $T(N)$ (Eager et al. 1989) (see Figure 2.7).

Note that each point in the graph corresponds to the execution of the application on a given number of N processors. The value of N increases as we go from right to left.

The completion time-efficiency profile for the example application characterized by Figure 2.1 and Table 2.1 is presented in Figure 2.8. Note that the $T(N)$ values used in the plot correspond to a PS scheduling discipline (discussed earlier) used during the execution of the application on a given number of processors.

2.5.4.1 The Knee

The completion time-efficiency profile aptly captures the trade-off between $E(N)$ and $T(N)$: as N is increased both $T(N)$ and $E(N)$ are decreased. That is, an improvement in completion time is accompanied by a decrease in efficiency. The value of N that leads to an effective compromise between these two metrics is intuitively a reasonable system operating point. This point of compromise is given by the values of N at which the ratio between $E(N)$ and $T(N)$ is maximized.

Figure 2.7 Completion time-efficiency profile.

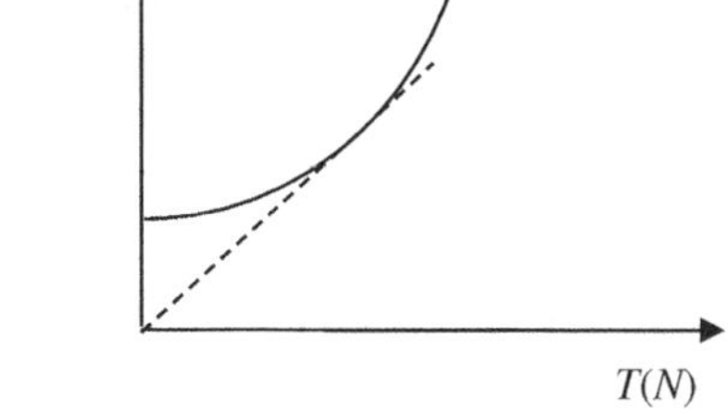

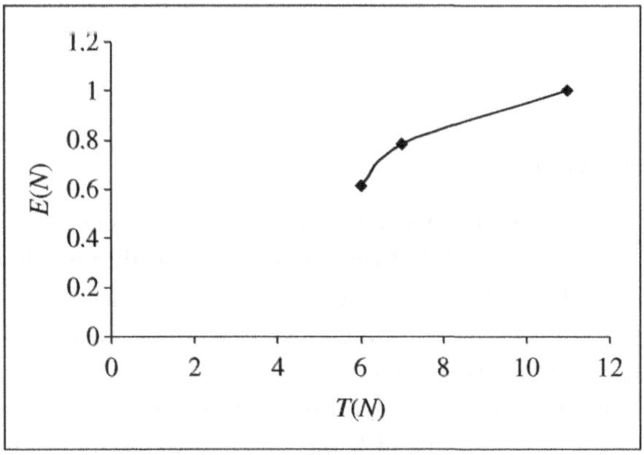

Figure 2.8 Completion time-efficiency profile for the example application.

2.6 Impact of Parallelism Characteristics on Performance

As discussed at the beginning of this chapter the parallelism characteristics of applications have an intimate relationship with system performance. Such relationships are explored further in this section. The impact of the fraction of sequential work, f, on speedup is presented in Section 2.6.1 while the relationship between average parallelism, A, and performance is discussed in Section 2.6.2.

2.6.1 Impact of Fraction of Sequential work

Amadhal's law captures the effect of f on speedup. Introduced by G. Amdahl in the late 1960s and described in (Bryant and O'Hallaron 2016), the law provides an upper bound on speedup, $S(N)$, achieved on an N-processor system:

$$S(N) \leq 1/\left[(1-f)/N + f\right].$$

The law follows from the observation that the part of the application that is strictly sequential leads to a speedup of 1, irrespective of the number of processors used, and from the assumption that at most a linear speedup is achieved for the remaining parts of the application.

Thus,

$$T(N) \geq f\,W + \left[(1-f)\;W/N\right], \tag{2.4}$$

where W is the total work (CPU seconds) resulting from thc cxccution of the application.

The first term on the right-hand side of the previous inequality corresponds to the time taken by the application to complete its sequential part, whereas

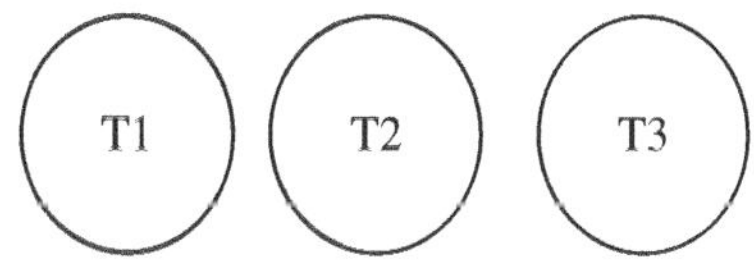

Figure 2.9 Example application.

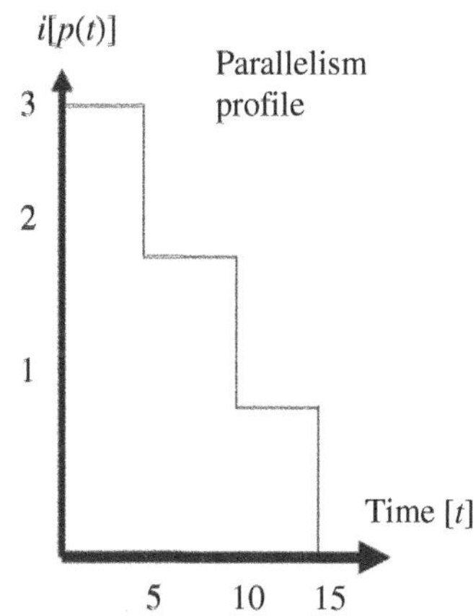

Figure 2.10 Parallelism profile for the example application of Figure 2.9.

the second component is based on the assumption that the remaining work is "perfectly" parallelizable leading to a linear speedup.

Dividing $T(1)$ by $T(N)$ and observing that $T(1) = W$ leads to the upper bound on speedup captured in the inequality preceding inequality (2.4).

An example of the bound on $S(N)$ based on f is presented next. The application comprises three independent tasks T1, T2, and T3 (see Figure 2.9) with execution times 5, 10, and 15, respectively.

Since none of the tasks has any predecessor, all of them start running when the number of processors is unbounded. So, for time 0–5, T1, T2, and T3 are running. After T1 completes at time = 5, T2 and T3 keep on running until time = 10. T2 completes at time = 10 whereas T3 at time = 15.

The parallelism profile for the application is presented in Figure 2.10.

From Figure 2.10 we get $p1 = 5/15$; $p2 = 5/15$; and $p3 = 5/15$. Note that $f = 5/15$.

These values can be used to obtain various values for $S(N)+$, the upper bound on speedup for a given value of N. A plot of $S(N)+$ for various values of N is provided in Figure 2.11.

2.6.2 Impact of Average Parallelism

The average parallelism of an application has an intimate relationship with speedup and efficiency. Given the average parallelism A, the speedup attained by the application for any given number of processors is bounded as (Eager

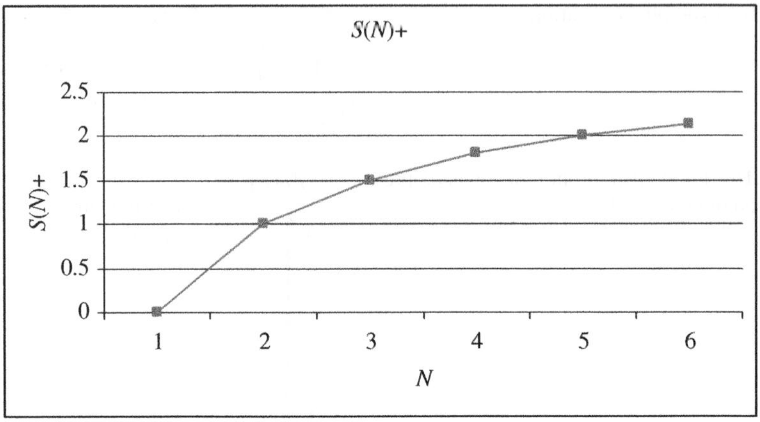

Figure 2.11 Upper bound on speedup based on f.

et al. 1989):

$$\min(N, A) \geq S(N) \geq NA/(N + A - 1)\,. \tag{2.5}$$

A discussion of the derivation of the inequality is beyond the scope of this chapter. A detailed derivation of the inequality can be found in the paper referenced earlier. A plot for $S(N)$ versus N is presented in Figure 2.12.

The bold line that corresponds to the minimum of N and A is the upper bound on speedup, whereas the (lower) dashed line is the lower bound on $S(N)$. Given A, the actual speedup curve of speedup for the application must lie between these two lines. Using the relationship between $S(N)$ and $E(N)$ leads to the following bound on efficiency:

$$\min(1, A/N) \geq E(N) \geq A/[N + A - 1]\,. \tag{2.6}$$

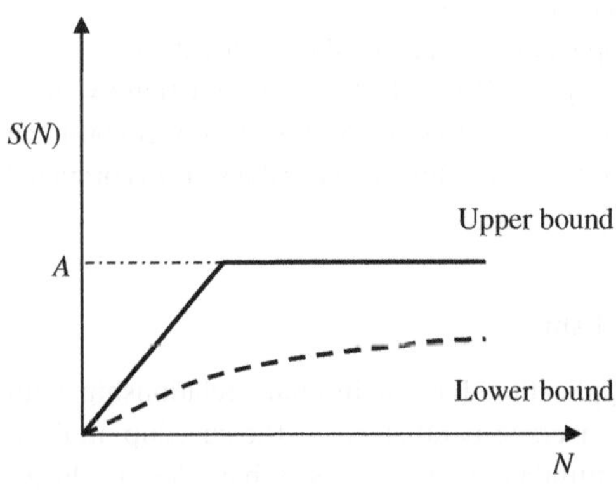

Figure 2.12 Bounds on speedup based on A.

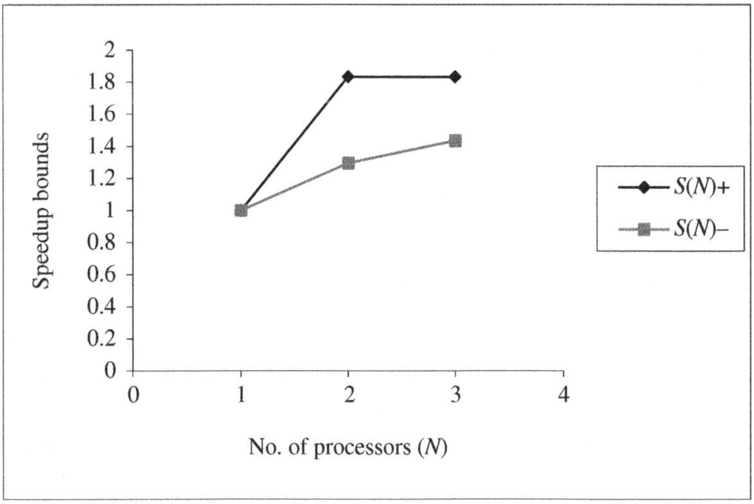

Figure 2.13 Bounds on speedup for example application.

The bounds for the example parallel application characterized by the precedence graph in Figure 2.1 and the task execution times presented in Table 2.1 are presented in Figure 2.13.

Note that the Average Parallelism, A, for the application, is 1.833 (see Section 2.4.3).

The upper bound and lower bound on the graph are labeled $S(N)+$ and $S(N)-$, respectively.

2.6.2.1 Bounds Based on Multiple Characteristics

Tighter bounds on speedup can be obtained by using multiple characteristics and the knowledge of the scheduling policy used for scheduling the tasks of the application on processors. The bounds on speedup, $S(N)$, by using A in conjunction with f when the PS scheduling discipline are described in (Eager et al. 1989):

$$\min[(N/(1+f(N-1)),A)\geq S(N)\geq NA/(N+A-1-(1-fA)).$$

The relationship between $S(N)$ and $E(N)$ can be used with the bounds on $S(N)$ to derive the bounds on $E(N)$.

2.7 Energy Performance Trade-Off

Energy consumed by large distributed systems such as cloud data centers used for both academic as well as enterprise computing typically consumes a great deal of energy. The energy cost of a data center is recognized to be an important component of its total operational cost. In addition to cost, the concomitant carbon footprint of the data center leads to greenhouse gas emission that is detrimental for the environment. Thus, power-conscious system deployment and operation are important concerns. A significant component of the energy-related cost is the cost related to the power consumed by servers and by the air-conditioning units that are necessary to cool these servers. A number of energy characteristics of applications and energy-related metrics are introduced in this section that focuses on the relationship among parallelism characteristics, energy characteristics, energy consumption, and performance.

Energy consumed by a system comprising multiple servers has an important relationship with its performance. Achieving higher performance often comes at the cost of increased energy consumption. The speed of a CPU can be controlled by its clock frequency; a higher clock frequency resulting in higher CPU speed leads to a higher power consumption by the CPU. Using multiple processors can also lead to speeding up the application completion time. But, using multiple-power-consuming processors can increase the energy consumption of the system.

Performance aware and energy aware system design are discussed in the literature. However, it is important to consider them together in many situations. By focusing on the relationship between performance and energy consumption this section outlines a number of energy characteristics and metrics for energy consumption. Using the relationship between such energy characteristics and the parallelism characteristics described in the previous sections, the trade-off between energy and performance is highlighted in this section. Bounds on energy consumption and performance are described.

The subsequent discussion is based on the following assumptions.

- Energy refers to the energy consumption by the CPU.
- The CPU power consumption remains constant when it is not idle.

2.7.1 Relationship Between Energy Consumption and Speedup

The speedup achieved by a parallel computation when executing on N processors is related to the energy consumption of the system. It is related to both the parallelism characteristics of the application discussed earlier in this chapter as well as the power consumption characteristic of the CPU. One such characteristic is *idle power ratio (b)*, which is the ratio of the power consumed by the CPU when it is idle and the power consumed by it when it is busy. Thus, the lower and upper bounds on b are 0 and 1: $0 \leq b \leq 1$. This idle power ratio is useful in understanding the relationship between energy consumed and performance that is discussed.

Given the value of b and the parallelism characteristics of an application it is possible to capture the relationship between application speedup and energy consumed through the energy factor that is defined next.

Energy Factor (P(N)): is defined as the ratio between the energy consumed on N processors and the energy consumed on a single processor for executing the application.

The relationship between $P(N)$ and speedup $S(N)$ for a given number of processors N is given by (Majumdar 2010):

$$P(N) = 1 - b + b\ (N/S(N)), \tag{2.7}$$

when the application has a linear speedup, $S(N) = N$, and $P(N)$ evaluates to be equal to 1. This is because the increase in energy consumption by using multiple processors is compensated by the decrease in the application completion time during which the processors are used.

Most real applications are typically characterized by a sub-linear speedup ($S(N) < N$), resulting in a $P(N)$ higher than 1. Higher the values of $P(N)$ higher is the energy consumption on an N processor system executing the application.

The equation characterizing $P(N)$ implies that $P(N)$ attains its highest value for the smallest value, $S(N)-$ for the application speedup. Similarly, $P(N)$ is minimized with the upper bound on speedup $S(N)+$.

Thus, the upper bound on $P(N)$ is related to $S(N)-$ as:

$$P(N)+ = 1 - b + (bN/S(N)-). \tag{2.8}$$

Similarly, the lower bound on $P(N)$ can be expressed as:

$$P(N)- = 1 - b + (bN/S(N)+). \tag{2.9}$$

An energy-conscious user may like to budget the energy usage on the system by specifying a budget on energy and estimate the speedup that can be attained on system that is operating within the constraints of the energy budget. An energy budget Pb is defined in (Majumdar 2010) as the ratio of the maximum energy that can be consumed on a parallel system comprising N processors and the energy consumed on a single-processor system.

The following inequality follows from the observation that the maximum energy consumed on an N processor system occurs when the lower bound on $S(N)$, $S(N)-$ is achieved on the system

$$S(N) \geq bN/(Pb + b - 1).$$

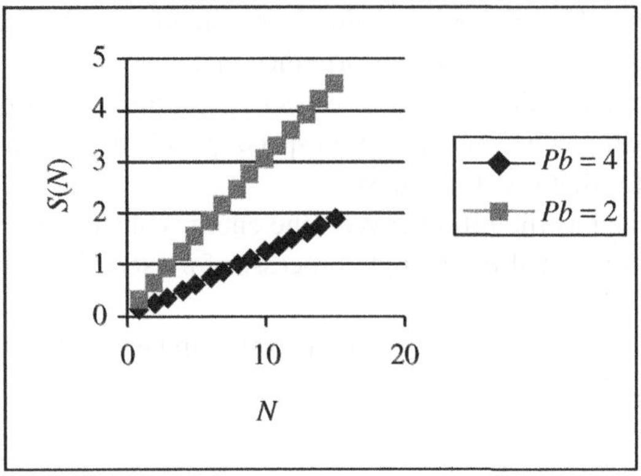

Figure 2.14 Relationship between energy budget and speedup.

This inequality can be rearranged to achieve the following new inequality:

$$Pb \geq b\ (1 + N/S\,(N)) - 1. \tag{2.10}$$

This relationship can be used to answer the question: what is the minimum value of the energy budget for achieving a given speedup?

An example usage of the relationship between *S*(*N*)– and *Pb* is captured in Figure 2.14 for two *Pb* values 2 and 4. Given a *Pb* selected by the user, the number of processors needs to be selected in such a way that the speedup for the application *S*(*N*) lies on or above the straight line corresponding to the specific *Pb* value.

2.7.2 Relationship Between Energy and Fraction of Sequential Work

As captured in Amdahl's Law discussed in Section 2.6.1 the fraction of sequential work of an application has an intimate relationship with the speedup *S*(*N*) achieved by the application. The relationship between speedup and energy factor leads to the following relationship between *P*(*N*) and *f* (Majumdar 2010):

$$P\,(N) \geq 1 + bf\ (N - 1).$$

Thus, *P*(*N*) increases with an increase in *b* or *f*. This is because an inefficient use of multiple processors resulting from a high idle power ratio *b* or from a large sequential component in the applications leads to an inefficient use of energy consumed by the system.

In addition to *f*, an energy budget, *Pb* can constrain the speedup attained on the system even further. This is captured in the following inequality (Majumdar 2010):

$$S\,(N) \leq \min\left(\frac{N}{1 + f\,(N-1)}, \frac{Pb - 1 + bf}{(b + Pb - 1)\,f}\right). \tag{2.11}$$

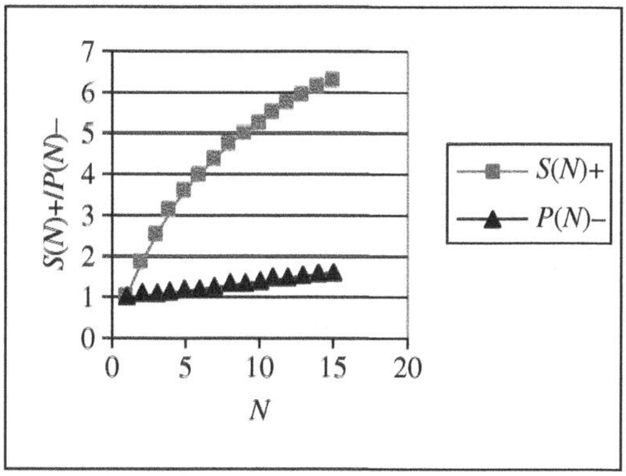

Figure 2.15 Bounds on $S(N)$ and $P(N)$. Source: Majumdar (2010)/with permission of Springer Nature.

The first term on the right-hand side of the inequality is a result of the software characteristics of the system (f), whereas the second term results from the value of f, b, the idle power ratio of the processors used and the power budget Pb imposed by the user.

Numeric Example:

The lower bound on $P(N)$, $P(N)-$ computed from Equation (2.9), and the upper bound on $S(N)+$ computed from the inequality (2.11) are plotted for $f = 0$ in Figure 2.15.

The value of b is 0.43, which corresponds to the example presented in (Majumdar 2010).

The figure shows that as N increases the upper bound on speedup is observed to increase rapidly at first but the rate of increase decreases at higher values of N. $P(N)-$ on the contrary is observed to increase steadily with an increase in N.

Energy Performance Trade-off: The improvement in speedup by using a high value of N may give rise to an increase in power consumption. The interaction between performance improvements and energy consumption is captured by the performance-energy ratio PE(N):

$$\mathrm{PE}(N) = S(N)/P(N).$$

Given N a high speedup or a small $P(N)$ results in a higher value of this ratio PE(N). The upper bound on PE(N) referred to as PE(N)+ is achieved for $S(N)+$ and $P(N)-$. Given f, the value of PE(N)+ can thus be computed from inequality (2.11) and Equation (2.9). PE(N)+ computed for two example use cases with $f = 0.1$ and $f = 0.5$ are presented in Figure 2.16.

The advantage of a lower value of f, the proportion of sequential work, is shown in Figure 2.16. For a given N, the use case with $f = 0.1$ gives rise to a higher value of PE(N)+ in comparison to the use case with $f = 0.5$.

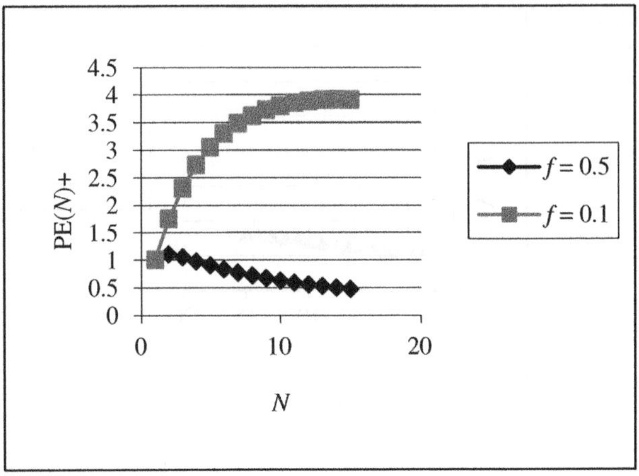

Figure 2.16 Effect of f on the upper bound on PE.

What is the highest value of $S(N)$ that can be achieved for a giver power budget is an important question. The metric $S(N) + Pb$ is used for this purpose and can be computed from inequality (2.11) for a given value of Pb and b. Figure 2.17 presents examples in which $S(N) + Pb$ is displayed for two different values of f. The power budget is shown to have a significant impact on $S(N) + Pb$ for the smaller value of f(0.1). Changing Pb from 1 to 2 brings about a large improvement in $S(N)$ + Pb that is observed to increase at a lower rate for further increases in Pb. The improvement

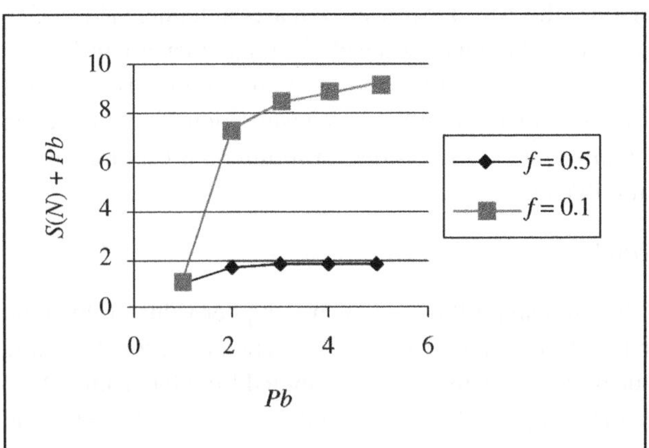

Figure 2.17 Relationship between power budget and speedup.

in $S(N) + Pb$ for various values of Pb is shown to be much more sluggish when the factor of sequential work is 0.5.

2.7.3 Relationship Between Energy and Average Parallelism

As discussed earlier, average parallelism A of and application has an important relationship with application speedup. Consequently, A is also related to the energy factor for an application. In fact, an upper and lower bound on $P(N)$ for a given number of processors N can be computed from the average parallelism of the application and idle power ratio:

$$1\text{-}b + b \max\left(1, \frac{N}{A}\right) \leq P(N) \leq 1\text{-}b + \frac{b(N + A - 1)}{A},$$

$P(N)-$ corresponds to the left-hand side of the inequality, and it follows from the expression for $P(N)-$ presented in Equation (2.9) and the upper bound on $S(N)$ based on A.

The upper bound $P(N)+$ is related to the lower bound of $S(N)$ and follows from using $S(N)-$ derived in inequality (2.5).

A numeric example for $A = 5$ and $b = 0.43$ is presented in Figure 2.18.

The upper and lower bounds on speedup and the energy factor are displayed in Figure 2.18. For $A = 5$ and various values of N. Both $S(N)+$ and $S(N)-$ are higher than the upper bound on $P(N)$. This implies that the speedup attained by an application for any given N is always higher than the respective energy factor. As N increases $S(N)+$ increases linearly at first resulting in a $P(N)-$ of 1 (see Eq. 2.9).

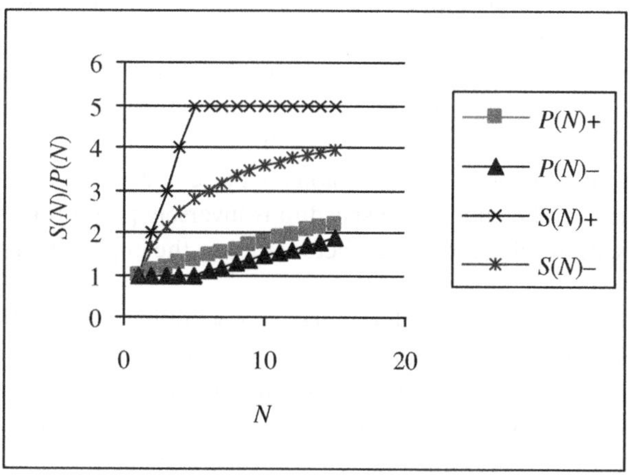

Figure 2.18 The impact of N on speedup and energy factor for a given A.

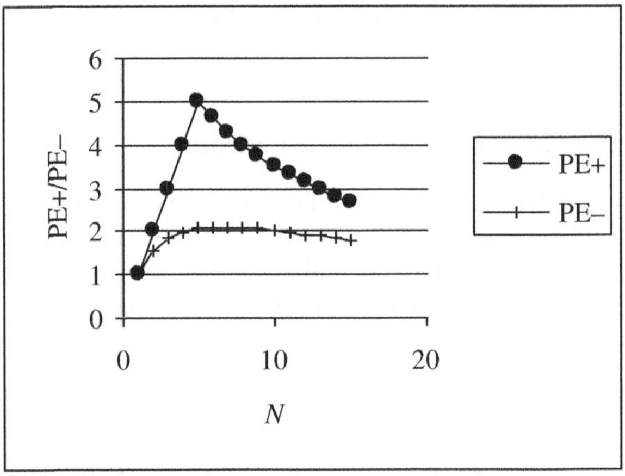

Figure 2.19 Bounds on PE for a given *A*.

After *S*(*N*)+ reaches its highest value, it levels off and *P*(*N*)– starts increasing linearly with *N*.

The trade-off between power and performance is captured by the performance-energy ratio. PE+ for a given *N* is the ratio between *S*(*N*)+ and *P*(*N*)– whereas PE– is the ratio between *S*(*N*)– and *P*(*N*)+ displayed in Figure 2.18. For $A = 5$ and a given *N*, both the upper and lower bounds on *S*(*N*) are larger than the upper and lower bounds on the energy factor, implying that *P*(*N*) is always lower than *S*(*N*) (see Figure 2.18). As *N* increases, *S*(*N*)+ increases linearly at first, resulting in a *P*(*N*)– of 1. After *S*(*N*)+ reaches the value of *A*, *S*(*N*)+ levels off and *P*(*N*)– starts increasing linearly with *N* (see Figure 2.18).

The increase in *S*(*N*)– with *N* follows inequality (2.5). The bounds on PE based on $A = 5$ for different values of *N* are plotted in Figure 2.19.

Both PE+ and PE– occur for $N = 5$ that is equal to the average parallelism of the application. The graph suggests using $N = A$ as a desirable operating point where the best trade-off between performance and energy is achieved.

Some key observations from the previous discussion are presented next.

Speedup and Energy Factor: The application speedup is inversely proportional to the energy factor. Thus, decreasing the energy factor requires the improvement of application speedup.

Performance-Energy Trade-off: The performance-energy ratio that aptly describes the energy-performance trade-off is observed to depend on both *f* and *A*. Both the upper and lower bounds on PE have a nonmonotonic relationship with *N*. When *A* is known using a values of $N = A$ at which both PE+ and PE– values are maximized suggest $N = A$ as a desirable system operating point.

Budget for Energy: The energy budget can have a significant impact on the variation in the upper bound on speedup with various values of *f*. A lower value of *f* leads to a higher *S*(*N*)+ for a given *Pb*.

2.8 Summary

Characterization of parallelism concerns the identification of those attributes of a parallel application that have a strong relationship with its execution behavior and performance. This chapter discusses a number of such characteristics of parallelism. Graph-based characteristics that include the parallelism profile, shape, and completion time-efficiency profile were described first. The single-point characteristics such as the average parallelism, knee of the execution time-efficiency profile, and fraction of sequential work were discussed later in this chapter. A number of performance metrics were described. The use of such characteristics in performance analysis and the relationship between such characteristics and the performance metrics were described. This discussion included well-known performance results that describe how to compute bounds on system performance using these characteristics.

Parallelizing applications and running them on multiple processors often comes at the cost of increased energy consumption. A number of chrematistics that capture the energy consumption resulting from the running of parallel applications on multiple processors are described. The combined relationship of these energy characteristics with parallelism characteristics such as fraction of sequential work and average parallelism and performance were discussed. This includes the description of how an energy budget specified by the user can influence system performance and power consumption.

Exercises

1 Consider the following precedence graph:

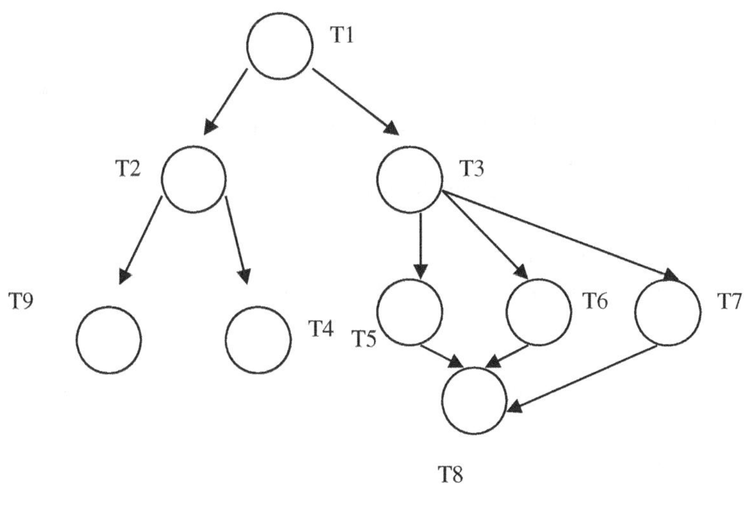

Process	T1	T2	T3	T4	T5	T6	T7	T8	T9
Execution time	5	2	3	7	3	2	1	5	3

(a) Draw the parallelism profile for the application
(b) Draw the shape for the application
(c) Plot the speedup as a function of the number of processors. Make sure that you also indicate the coordinates of each point in the plot in a separate table. For the computation of speedup, assume that the "Processor Sharing" discipline is used for process scheduling.

2 Consider a parallel application comprising multiple processes.
Given: $A = 5$. The number of processors on the system is N
Determine the upper and lower bounds on speedup attained by the application for $N = 1, 2, 3, 4, 5, 6$
Present your results in the form of a graph.

3 Consider a parallel application comprising multiple processes.
Given: $f = 0.1$. The number of processors on the system is N
Determine the upper bound on speedup attained by the application for $N = 1, 2, 3, 4, 5, 6$
Present your results in the form of a graph.

4 Consider a parallel application comprising 10 tasks: T1–T10. The execution times of all the tasks are the same. Upon start T1 starts executing. On completion of T1, T2–T10 become ready to run. Determine (a) the average parallelism for the application (b) fraction of sequential work (c) the maximum speedup that can be attained by the application.

5 Consider the Parallelism Profile (for a parallel application) with the following measurement data:
T_i = the total time during which the degree of parallelism is exactly equal to i $(i = 1 \ldots m)$
m = maximum parallelism for the application.
Derive an upper bound on speedup for the application using T_i, m, and N (the number of processors).

6 Consider n applications: App_1 ... App_n. App_i ($i = 1 \ldots n$) each of which is characterized by an average parallelism of A_i and a $T(\infty)$ of T_i. You have to consider a complex application that is composed of all the n applications.

Determine the average parallelism of the complex application when the n component applications App_1 … App_n are to be executed in parallel.

7 Consider the n applications described in Question 6.
Determine the average parallelism of the complex application when the n component applications App_1 … App_n are to be executed in sequence.

8 Consider two Applications App 1 and App 2 with exactly the same shape function. That is, value of P_{i1} for App 1 and any given i is the same as P_{i2} for App 2 and the same value of i. If $T(\infty)$ for App 1 is K times ($K > 0$) the $T(\infty)$ for App 2 determine the ratio between the completion time of App 1 and App 2 achieved for any given number of processors N and a Processor Sharing scheduling discipline.

References

R.E. Bryant, D. O'Hallaron. *Computer Systems: A Programmer's Perspective* (3 ed.), Pearson Education, London, UK, 2016, p. 58, ISBN 978-1-488-67207.

D.L. Eager, J. Zahorjan, E.D. Lazowska, "Speedup versus Efficiency in Parallel Systems", *IEEE Transactions on Computers*, Vol. 38, Issue 3, March 1989, pp. 408–423.

S. Majumdar, "On the Energy-Performance Tradeoff for Parallel Applications", *Proc. EPEW 2010: 7th European Performance Engineering Workshop*, Bertinoro, Italy, September 2010.

K.C. Sevcik, "Characterizations of Parallelism in Applications and Their Use in Scheduling", *Proc. ACM SIGMETRICS Conference*, May 1989, pp. 171–180.

3

Resource Management Techniques for Distributed Computing Systems

This chapter considers two types of systems discussed in Chapter 1: Shared memory systems and distributed memory systems. Each of these systems is characterized by multiple processing resources and are subjected to a workload that comprises jobs each of which consists of multiple concurrent work units introduced in Chapter 1. We will refer to such work units as processes in this chapter. However, the techniques discussed, if not specified otherwise, can be applied to both processes and threads.

Resource management is often said to comprise two techniques: allocation and scheduling. *Allocation* refers to the mapping of processes to processing resources. An allocation algorithm determines which process in an application will be executing on which of the resources in the system. *Scheduling* concerns the order in which the various processes running on the system are executed. For example, given a processor running a given set of processes, the scheduling algorithm needs to determine the start time for each of these processes. Allocation is discussed first in Section 3.1. Both well-known algorithms as well as guidelines for the devising of heuristic algorithms are described in the different subsections of this section. Scheduling is the subject of attention in the following sections. Scheduling on grids and clouds is discussed in Section 3.2. Special attention is paid to systems processing requests characterized by SLAs. Both algorithms and the impact of various system and workload parameters on the different resource scheduling algorithms are discussed. Allocation and scheduling are important in the context of various types of distributed systems. Resource management on grids and clouds processing requests characterized by deadlines is presented in Section 3.3. Scheduling on client–server systems, in which client tasks utilize the services of lower-level server tasks, is discussed in Section 3.4. A description of the software bottlenecking concept and how it can affect the performance of resource management on client–server systems is included. A summary of the discussions presented in this chapter is provided in Section 3.5.

Resource Management on Distributed Systems: Principles and Techniques,
First Edition. Shikharesh Majumdar.

3.1 Resource Allocation

Resource allocation is a computationally hard problem. The discussion is motivated with the help of an example that presents the well-known Graham's anomaly that was pointed out in Graham (1966).

3.1.1 Graham's Anomaly

Consider the precedence graph for a concurrent application that is executing on a three-processor system.

Π = {P1, P2, P3, P4, P5, P6, P7, P8, P9, P10}; {P = {(P1, P2), (P1, P3), (P1, P4), (P1, P5), (P2, P6), (P5, P6), (P5, P7), (P5, P8), (P5, P9)}. The precedence graph is displayed in Figure 3.1. The execution time of each process is presented in Table 3.1.

Consider an allocation algorithm in which processes are allocated on a first-come-first-served (FCFS) basis. Thus, as and when a process becomes ready to run (i.e. when all its predecessors have completed), it is inserted in a FIFO queue of ready processes. Whenever a resource becomes free, the first process at the head of the ready queue is allocated to the resource. Ties are broken by using the process index. That is, if Pi and Pi+k become ready to run at the same time Pi is chosen for execution first. The mapping of the processes and their start and completion times are captured in the Gantt chart presented in Figure 3.2. Note that between times 2 and 4 Processors 3 remains idle because there is no

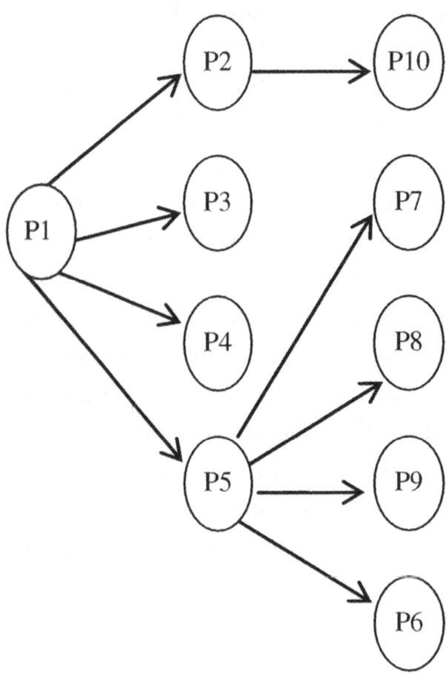

Figure 3.1 Example precedence graph.

Table 3.1 Process statistics (three processors).

Process	Execution time	Start time	Completion time
P1	2	0	2
P2	3	2	5
P3	2	2	4
P4	2	2	4
P5	2	4	6
P6	4	6	10
P7	4	6	10
P8	4	10	14
P9	4	10	14
P10	9	5	14

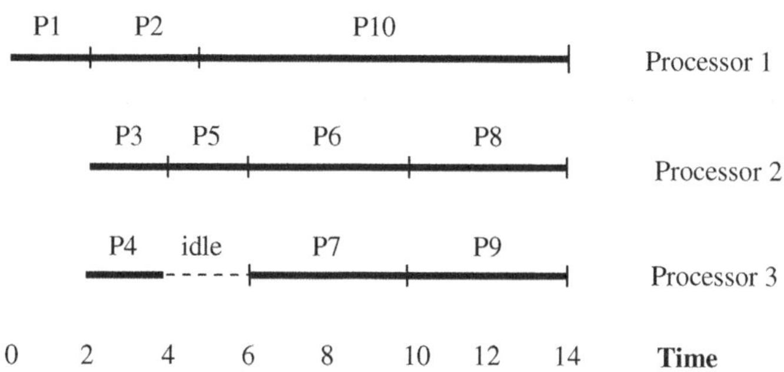

Figure 3.2 Gantt chart for example precedence graph presented in Figure 3.1 (number of processors is three).

ready process during that time in the ready queue. The application completes at time = 14 when all its processes are completed.

Now, consider the execution of the same application on four processors. The Gantt chart is provided in Figure 3.3 and the process statistics in Table 3.2.

Instead of improving system performance, the increase in the number of processing resources increases the completion time of the application to 17. This anomalous behavior was reported by Graham, and is thus referred to as *Graham's Anomaly*.

3.1.2 The Impact of Processor Allocation on Performance

The mapping of processes to processors for a given number of processors can have a significant impact on performance. The relationship between the number

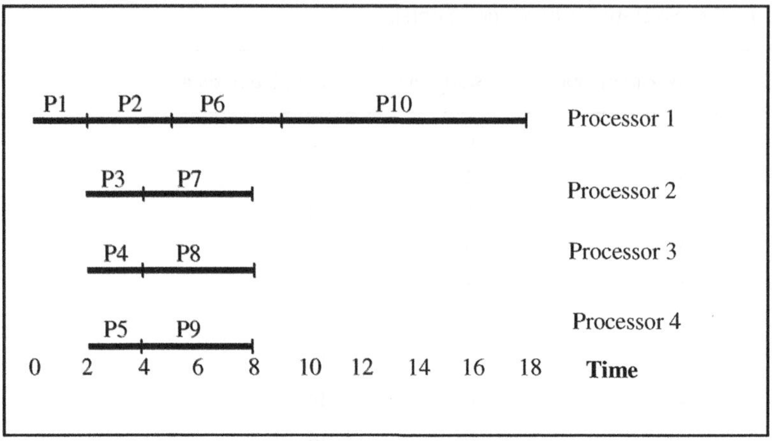

Figure 3.3 Gantt chart for example precedence graph presented in Figure 3.1 (number of processors is four).

Table 3.2 Process statistics (four processors).

Process	Execution time	Start time	Completion time
P1	2	0	2
P2	2	2	5
P3	2	2	4
P4	2	2	4
P5	2	2	4
P6	4	5	9
P7	4	4	8
P8	4	4	8
P9	4	4	8
P10	9	9	18

of processors and the mapping of processes to processors on completion time achieved by an application is captured in (Graham March 1969):

$$T(N')/T(N) \leq 1 + (N-1)/N'.$$

The left-hand side is the ratio of the completion time $T(N')$ achieved when the application is run on N' processors and the completion time $T(N)$ achieved when the application is run on N processors. Note that the allocation policies used with N and N' are not constrained to be the same. If we set $N' = N$, then the ratio shows the ratio between the largest possible value of $T(N)$ achieved with the worst mapping of processes to processors and the smallest possible value of $T(N)$ achieved with the best mapping of processes to processors. Thus,

$$T\left(N, \text{worst mapping}\right)/T\left(N, \text{best mapping}\right) \leq 2\text{–}1/N.$$

3.1.3 Optimal Allocation Techniques

Processor allocation to applications belongs to the class of NP-hard problems. The time for finding the optimal solution for such allocation is observed to grow rapidly with the size of the problem. Consider, for example, the allocation of the processes in a precedence graph on N processors. To find the mapping of processes to processors that gives rise to the smallest application completion time one may need to find out the application completion time for each possible allocation of processes to processors. The number of such combinations increases exponentially with the number of processes and processors in the application, and the time to find the optimal solution can become very large quickly with an increase in the number of processes in the application.

3.1.3.1 Hu's Algorithm

Although processor allocation is an NP-hard problem in general, there are specific classes of precedence graphs for which the optimal solution can be found in polynomial time. Such an algorithm proposed by Hu (1961) is described. The algorithm was proposed for scheduling work among multiple men and is adapted to the context of scheduling processes on multiple processors.

Hu's algorithm can be used only with tree-structured precedence graphs in which all processes have the same execution time. This graph is characterized by a root node and a number of predecessor nodes organized in the form of a tree. An example of such a graph is shown in Figure 3.4.

If the processes in an application do not have the same execution time, in some cases, the precedence graph can be restructured, with some processes split into multiple smaller processes in such a way that each process in the restructured application has the same execution time. An example of such a restructuring is shown. Figure 3.5 presents the original precedence graph and Figure 3.6 the restructured precedence graph. The execution times of P1, P2, and P3 in the figure are 2, 1, and 1. Process P1 is split into two processes, P11 and P12, each with an execution time of 1. As a result, all the processes in the restructured precedence graph have the same execution time of 1. Note that such a transformation may not be possible if the application design does not allow the splitting of process P1.

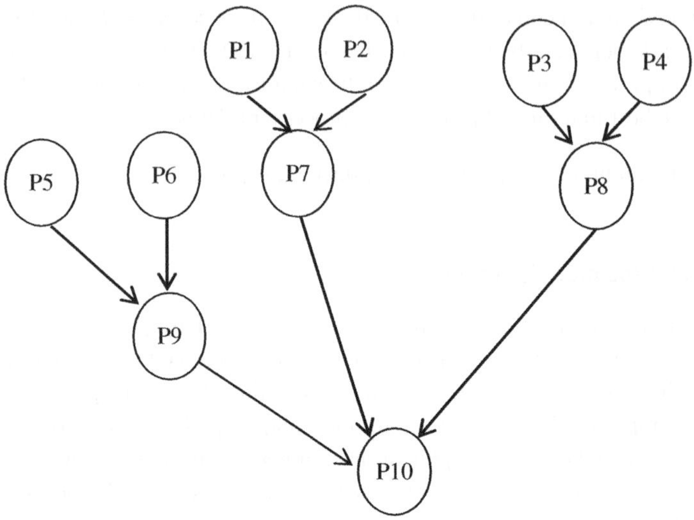

Figure 3.4 Example tree-structured precedence graph.

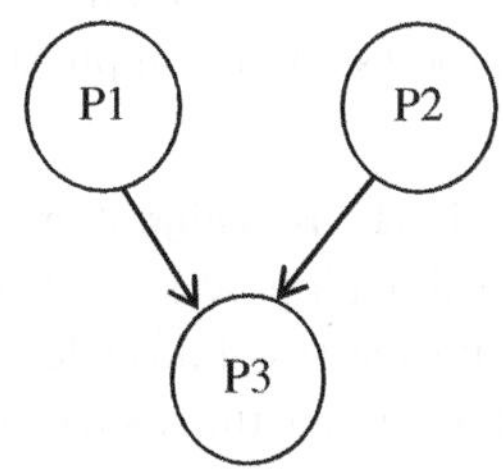

Figure 3.5 Original precedence graph.

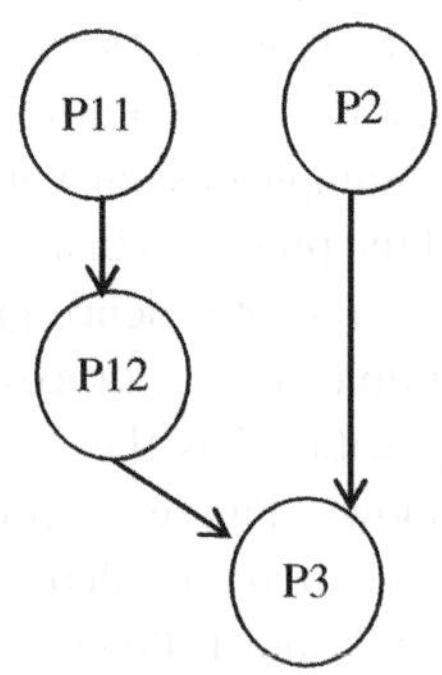

Figure 3.6 Restructured precedence graph.

Before presenting Hu's algorithm, the notion of a path and its length is introduced. A path from a given node, *p*, comprises the set of nodes traversed to reach the root node from *p*. The length of this path, length (*p*), is defined as the number of edges that need to be traversed to reach the root node (e.g. process P3 in Figures 3.5 and 3.6) from *p*).

Hu's algorithm is presented in Figure 3.7.

It is an iterative algorithm that works until all the processes in the application are allocated to processors. An example tree-structured precedence graph with 10 processes is presented in Figure 3.8. Consider a system comprising three processors. The resultant processor allocation achieved with Hu's algorithm at various points of time during algorithm execution is shown in Figure 3.9. Note that all process execution times are the same. So, the duration of each time step in Figure 3.9 is equal to this common process execution time.

Upon start (line 1), all the processors are found to be idle and line 2 is executed where in *R* is assigned the set of all ready processes {P1, P2, P3, P4, P5, P6}. Since the path lengths of P1, P2, P3, and P4 are the highest,

1. Wait for a processor to become idle.
2. R = set of all processes for each of which all its predecessors have completed.
3. Compute R', a subset of R, such that p is in R' if length (p) is maximal.
4. Choose an arbitrary p in R' and assign it to an idle processor.
5. If there are no other processes to schedule, then stop;
   Otherwise, go to step 1.

Figure 3.7 Hu's algorithm.

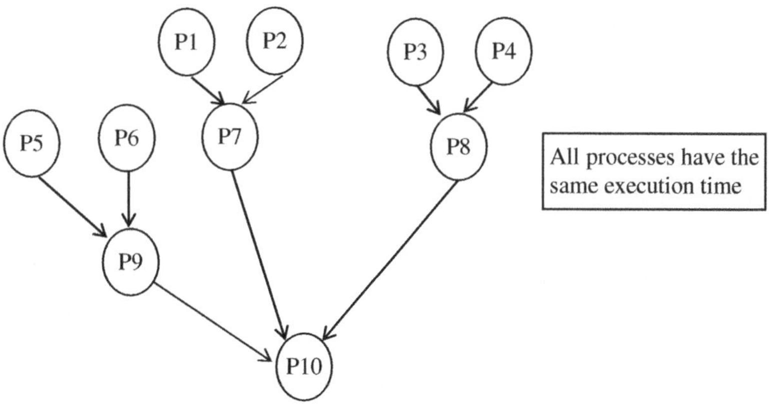

Figure 3.8 Example application with tree-structured precedence graph.

Processor	Process allocated (time step: 0–1)	Process allocated (time step: 1–2)	Process allocated (time step: 2–3)	Process allocated (time step: 3–4)
Processor 1	P1	P2	P7	P10
Processor 2	P3	P5	P8	
Processor 3	P4	P6	P9	

Figure 3.9 Allocation of processes to processors for Hu's algorithm.

R' is determined as the set {P1, P2, P3, P4} in line 3. Since the number of processors is smaller than the cardinality of R' any arbitrary processes in a subset of R' (P1, P3, and P4 in this example) are chosen to execute on the three processors (line 4). In line 5, since there are processes remaining to be allocated the algorithm loops back to step 1 and waits until P1, P3, and P4 that are executing on the systems during time steps 0–1 are completed. The algorithm then executes line 2. Note that after P3 and P4 are complete, their successor P8 becomes ready to run. Thus, the value of R computed in line 2 is {P2, P5, P6, P8}. Out of these (P2) has the largest path length, and R' is set to P2 (line 3) and P2 is allocated on processor 1 (line 4). In line 5, the algorithm loops back to line 1, and since there are two idle processors, it goes to line 2 and computes R' as {P5, P6, P8} all of which have the same path length of 2 (line 3). Any arbitrary set of two processes (P5 and P6 in this example) are assigned on processor 2 and processor 3 (line 4). Since there are more processes that remain to be allocated the algorithm, in line 5, the algorithm loops back to line 1 and waits until time steps 1–2 are completed. Note that during this time, steps P2, P5, and P6 are executed in the system.

The algorithm continues in this way until all the processes in the application are allocated. Note that as shown in Figure 3.9 the application is completed at the end of time steps 3–4.

3.1.4 Heuristic Techniques for Processor Allocation

Since optimal solutions are not always possible to find in a reasonable period of time, heuristic techniques are often used. Although such a technique may not be able to determine the optimal solution, it is able to find a reasonable sub-optimal allocation that can be computed in a reasonable period of time. Three different principles that can be used for devising such heuristics are described next. This is followed by descriptions of two well-known heuristic algorithms.

Allocate Bottleneck Processes First

A bottleneck process is the process that has the strongest impact on completion time and needs to be considered for allocation first. Consider, for example, an application comprising four independent tasks P1, P2, P3, and P4 shown in Figure 3.10. The bottleneck process is the process with the largest processing time: P3. Consider an execution of the application on two processors. P3 needs to be allocated first and is placed first on processor 1. The other processes are allocated to processor 2. This will lead to an application completion time of 10, the smallest completion time when the application is run on two processors. One such deployment is captured in Table 3.3, which shows the execution time and completion times of the four processes.

The application completes at time = 10 when all its processes are complete. In this example, the completion time corresponds to the completion time of the longest (bottleneck) process P3.

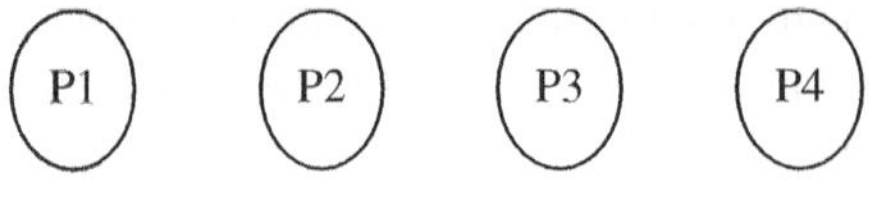

Figure 3.10 Example application for demonstrating the application of the allocate bottleneck processes first.

Table 3.3 Timing data for the execution of application shown in Figure 3.10 on two processors.

Process	Execution time	Allocated on	Start time	Completion time
P1	1	Processor 2	8	9
P2	3	Processor 2	5	8
P3	10	Processor 1	0	10
P4	5	Processor 2	0	5

Table 3.4 Timing data for the execution of the application is shown in Figure 3.10 achieved with an alternate allocation on two processors.

Process	Execution time	Allocated on	Start time	Completion time
P1	1	Processor 1	0	1
P2	3	Processor 1	1	4
P3	10	Processor 1	4	14
P4	5	Processor 2	0	5

Table 3.5 Processor load data for the execution of the application shown in Figure 3.10 on two processors for an allocation that is based on the balancing of the load principle.

Processor	Processes allocated	Load
Processor 1	P1, P2, P4	9
Processor 2	P3	10

Table 3.4 presents the timing data when a different allocation for which the allocation of P3 was delayed, and it was the third task to execute on processor 1. This allocation leads to a higher application completion time of 14.

Balance Load across the Processors

This principle advocates the allocation of processes to processors in such a way that the load is spread evenly across all the processors. Consider the execution of the previously described application on two processors. Table 3.5 shows such a balanced allocation for which 9 units of work is allocated on Processor 1 and 10 units of work on Processor 2.

The application completes at time = 10 when all the processors have finished executing the processes allocated to it. For a different allocation captured in Table 3.6, the application completes at time = 15. It can be observed that for this

Table 3.6 Processor load data for the execution of the application shown in Figure 3.10 on two processors for an allocation that leads to an unbalanced processing load.

Processor	Processes allocated load	Load
Processor 1	P1, P2	4
Processor 2	P3, P4	15

allocation the load is not balanced. Processor 1 is allocated 15 units of work, whereas processor 2 is allocated 4 units of work.

Reduce Interprocess Communication

Processes in an application often communicate with one another. Interprocess communication takes longer if the communicating processes are running on different processors. This can increase the completion time of the application. Thus, this principle advocates the co-allocation of the processes that intercommunicate frequently on the same processor. Note that, this can give rise to a conflict with the load-balancing principle discussed earlier. It is up to the designer to decide which principle will be better to follow in a given situation.

3.1.4.1 Largest Processing Time First

This algorithm is based on the largest processing time (LPT) rule. The LPT rule considers k jobs to be executed on N processors in such a way that the makespan achieved in the system is minimized. The makespan is equal to the difference in time at which the last job completes execution and the start time of the first job. The LPT rule advocates the ordering of jobs in a nonincreasing order of execution times. The jobs are then assigned one at a time starting with the job at the head of the list that is allocated on the processor with the smallest load so far. It was shown that the makespan, T(LPT) achieved with LPT on N processors is bounded as (Graham March 1969):

$$T(LPT) \leq 4/3 - 1/3N.$$

The largest processing time first (LPTF) algorithm that is based on the LPT rule and is described in Figure 3.11. From the description, it is clear that the algorithm

Step 1: Order the ready processes in nonincreasing order of execution times
Step 2: If a processor is idle allocate the process at the head of the list to the processor
Step 3: If all tasks are allocated stop
Else wait until there is at least one ready process and at least one idle processor on the system. Then go to step 1.

Figure 3.11 LPT rule.

allocates the largest tasks first and is thus in line with the "allocate bottleneck process first" principle. By avoiding the placing of ready tasks on a busy processor, it also complies with the "load-balancing" principle.

3.1.4.2 The Multifit Technique

An allocation framework called Multifit (Coffman et al. 1978) that is used for allocating tasks in an application consisting of multiple independent processes on a system comprising multiple processors is described. Since the processes are independent, the application is assumed to be free from any intertask communication. Multifit comprises two components.

The first concerns the ordering of processes for allocation, and the second concerns the selection of the processor on which the next task is to be allocated. The specific criterion used by each component is presented.

Ordering Criterion: specifies the order in which the processes in the application are to be organized for allocation. Examples include random order and nondecreasing order of execution time.

Placement Criterion: focuses on determining which processor the next task will be allocated on. Examples include First Fit and Best Fit, which will be explained later.

A bin-packing model for processor allocation is used. Each bin represents a processor and each object a process that needs to be placed in a bin.

An example of the multifit approach when the ordering criterion for processes is *nonincreasing* order of process execution times, and the fitness criterion is *first fit* is presented in Figure 3.12. It can be easily adapted to other ordering and fitness criteria.

The system has m processors, and the number of processes to be allocated is N.

In step 1, the N processes are sorted in nonincreasing order of process execution times and are stored in the variable list. Step 2 computes Cmin and Cmax, which

```
Step 1. Sort processes in accordance with ordering policy and store in list
Step 2: Calculate capacity bound: Cmin and Cmax
Repeat
        Step 3. Capacity C = (Cmin + Cmax)/2
        Step 4. For i = 1 to N
                Choose ith process from list;
                Place it on the first bin (processor) that can accommodate the process,
                creating a new bin for placing the ith process if none of the existing bins
                has a leftover capacity to accommodate the ith process;
        Step 5. If no of bins used > m {Cmin = C}
                else {Cmax = C}
until the maximum number of iterations is reached.
```

Figure 3.12 The multifit approach.

are used in the following step of the algorithm. A discussion of possible ways for computing Cmin and Cmax is provided later. An iteration spanning steps 3–5 is described next. Each bin is assigned a capacity of C that is computed in step 3. In step 4, which is executed N times, the ith process is placed in a bin using the first fit criterion. That is, all the bins are scanned from the beginning and a bin with enough leftover capacity to accommodate the ith process is chosen. If none of the bins can accommodate the process, a new bin is created to accommodate the process, and the list is modified appropriately. If the number of bins used is greater than the number of processors m (step 5), the solution is deemed infeasible, and the capacity C to be used in the next iteration is increased by setting Cmin to the current value of C. Otherwise, the value of C to be used in the next iteration is decreased by setting Cmax to the current value of C.

The loop (steps 3–5) is repeated until the maximum number of iterations is reached. The mapping of processes (objects) to processors (bins) is always stored at the end of a given iteration. Thus, when the algorithm terminates the mapping generated in the last iteration becomes the output of the algorithm. After termination of the algorithm if the number of bins used is more than the number of processors the user has the choice of repeating the execution of the algorithm for different values of Cmin, Cmax and the maximum number of iterations.

Cmin and Cmax are expected to be specified by the user. One possible value for Cmax is the sum of the execution times of all processes. Clearly, the completion time with a given number of m processors cannot exceed this sequential execution time for the application. A possible value for Cmin is the execution time of the largest process in the application. This is the minimum value for completion time $T(\infty)$ achieved by running the application on an unbounded number of processors. Other methods of computing the values of Cmin and Cmax are available in the literature.

3.2 Task/Process Scheduling

Scheduling is another important component of resource management. This section focuses on the scheduling of multiple tasks/processes on a processor. Such an operation may need to be performed after the allocation of tasks to processors. The scheduling algorithm needs to determine the start time for each process in the process set allocated on a processor so that the service level agreements (SLAs) are met. The words task and process will be used interchangeably in the chapter. Since the unit of work (process/task) is often associated with a user request that arrives on the system, we also use the term request to refer to the unit or work.

3.2.1 Scheduling of Ready Tasks on a Single Processor

This subsection focuses on the scheduling of a set of tasks waiting to execute on a given processor. Some of the well-known techniques used in conventional multiprogrammed systems are FCFS, Shortest Job First, Shortest Remaining Time First, and Round Robin, which are discussed in many operating systems textbooks, such as (Silberschatz et al. 2014). Additional scheduling policies for such systems that include multiple queue-based techniques are also described in such books. This book discusses the scheduling of requests associated with SLAs that are often used on clouds. The SLA parameters include (i) the earliest time of execution for the task, (ii) task execution time and (iii) a deadline by which the task must complete. Requests associated with such tasks are also referred to as advance reservation requests (ARs) (Foster et al. May 1999) because for the acceptance of the request a cloud service provider needs to reserve a resource for the task such that the SLA parameters are met. A request that does not specify any earliest task execution time or deadline is often referred to as an on demand (OD) request and is served on a best-effort basis by the system.

3.2.1.1 Scheduling Tasks with SLAs

Scheduling a set of tasks each associated with its own SLA is discussed. The remainder of Section 3.2 focuses on systems with a single-processing resource, whereas Section 3.3 describes scheduling on systems with multiple-processing nodes.

A formal definition of the general scheduling problem is presented first.

Given a resource b and a set of tasks {i, j, …, k} with a set of earliest start times $\{t_i, t_j, \ldots, t_k\}$*, service times* $\{e_{ib}, e_{jb}, \ldots, e_{kb}\}$ *and deadlines* $\{d_i, d_j, \ldots, d_k\}$*, generate a schedule such that each task i starts executing after its earliest start time* t_i *and finishes before its deadline* d_i (Farooq et al. 2005a).

Like processor allocation, scheduling of tasks with SLAs is also an NP-hard problem. Finding the optimal solution requires exploring the entire search space, which can be very time-consuming when the number of processes involved is high. An algorithm *called Scaling through Subset Scheduling (SSS)* is presented in this section. Only a high-level overview of the techniques is presented. The interested reader can find the detailed algorithm in Farooq et al. (2005a). The technique is aimed at eliminating a set of tasks that are not relevant at a given point in time during the execution of the algorithm such that the search space for the solution is reduced.

Basic Approach: Whenever a new request/task arrives, the algorithm tries to find out whether or not the SLA associated with the task can be satisfied and constructs a new schedule that accommodates the newly arrived task while ensuring that the SLAs associated with all the tasks that have already been accepted by the system are not violated. It first identifies the set of all those tasks in the resource schedule that can affect the feasibility of the new schedule with the new request and then tries to work out a feasible schedule for only that subset of tasks S. Consider, for example, the arrival of a task m with an earliest start time s_m and a deadline of d_m. Those existing tasks in the system whose start times and deadlines do not overlap with the time period s_m to d_m cannot interfere with the scheduling of the newly arrived tasks and are therefore not included in S. Only those tasks that can be active during the execution of the newly arrived task are included in S. This reduces the number of tasks to be considered and the run time of the algorithm. A high-level description of the approach used in SSS is presented next.

- **Step 1:** Compute S that is the set of all those tasks, which can affect the scheduled time of the new task and whose scheduled time can be affected by the new task. The scheduled time of a task is the time at which the task is to start executing on the system.
- **Step 2:** Find an initial feasible schedule for tasks in S using a well-known heuristic, such as the earliest deadline first (EDF) strategy. EDF is a well-known strategy that associates a higher priority with tasks with shorter deadlines (Silberschatz et al. 2014).
- **Step 3:** If the schedule is feasible, accept the task and update the resource schedule. Otherwise, compute lower bounds on the lateness of the *critical task* and check if its lateness can be improved. The critical task is the task that among all the tasks considered misses its deadline by the largest amount. If it cannot be improved reject the new task (request). Otherwise, go to step 4.
- **Step 4:** Perform improvements on the initial schedule iteratively. The algorithm stops if a feasible schedule is found or the algorithm determines that no feasible solution exists.

A detailed discussion of the algorithm is provided in the paper referenced earlier.

3.2.1.2 Performance Analysis

A simulation-based performance analysis of SSS on an open system that is subjected to an open stream of AR and OD request arrivals described in Farooq et al. (2005b) is presented. The relationship of two important performance metrics and workload parameters is presented next. The performance metrics discussed are:

Blocking Probability (P_b): is the proportion of tasks that are rejected by the cloud service provider because its SLA parameters cannot be satisfied,

Average Response Time of OD Requests (R_{OD}): is the averages value of the difference between the time of completion and the time of arrival of the OD requests.

Since OD requests are satisfied on a best-effort basis an OD request is never rejected but the request may experience longer latencies in comparison to AR requests and R_{OD} is, therefore, an important metric to observe.

Resource Utilization (U): is the proportion of time the resource is utilized.

The paper discusses additional performance metrics but they are not relevant in the context of the discussion presented in this section.

Two important workload parameters are used in the discussion presented in this section.

Percentage of Advance Reservation Request (PAR) is the proportion of AR requests in the workload that comprise both AR as well as OD requests.

The *Laxity (L)* parameter represents the average proportion of slack time in requests. The laxity (slack time) of a given request is the difference between its deadline and the sum of the earliest start time and execution time. The percentage laxity of requests is the percentage value for the ratio between the laxity and execution time of the requests in the workload.

The Effect of PAR and L on P_b

Figure 3.13 presents the impact of PAR and L on P_b. As PAR increases, P_b is observed to increase. At PAR = 0, $P_b = 0$ because all the requests are ODs that are not associated with any deadlines and are therefore accepted. As PAR increases resource contention increases and the probability that a specific request's deadline cannot be met increases. As a result, P_b increases. An increase in laxity increases the average task slack time, and therefore, improves the possibility of meeting task deadlines. Thus, for a given PAR as L increases P_b decreases.

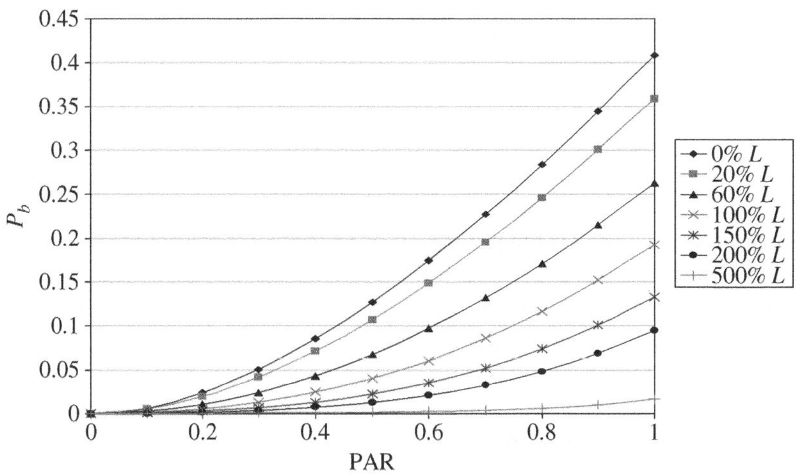

Figure 3.13 Effect of L and PAR on blocking ratio. Source: Adapted from Farooq et al. (2005b).

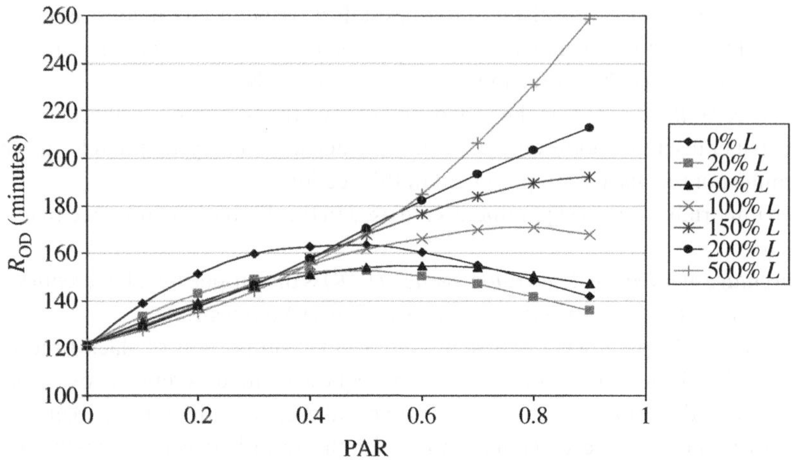

Figure 3.14 Effect of *L* and PAR on response time of OD requests. Source: Adapted from Farooq et al. (2005b).

The Effect of PAR and L on R_{OD}

The acceptance of an AR may lead to the postponement of the execution of an OD because the AR needs the resource at a specific point in time. As a result, as PAR increases R_{OD} is observed to increase (see Figure 3.14). It is interesting to note the nonmonotonic relationship between R_{OD} and PAR. For lower values of L, ROD increases with PAR initially (see for example, the graph for $L = 0$) but starts decreasing when PAR is increased further beyond a certain point. Describing the possible reason for this is left as an exercise for the reader.

The Effect of PAR and L on U

Resource utilization depends on both request arrival rate λ and blocking probability

$$U = \lambda^*(1 - P_b) * (R - W).$$

The product $\lambda^*(1 - P_b)$ is the proportion of requests that are accepted and, are therefore, executed on the resource. R is the mean task residence time and W is the mean task waiting time. Thus, $R - W$ is the mean task execution time. The right-hand side of the previous equation is thus the proportion of time the resource is busy executing the accepted tasks.

As PAR increases, the proportion of ARs arriving on the system increases. This increases system contention and P_b increases (see Figure 3.13). As a result, the number of requests accepted by the system decreases which seem to decrease resource utilization (see Figure 3.15). As laxity L increases for a given PAR, the ARs get more slack time and P_b decreases (see Figure 3.13). This is an indication

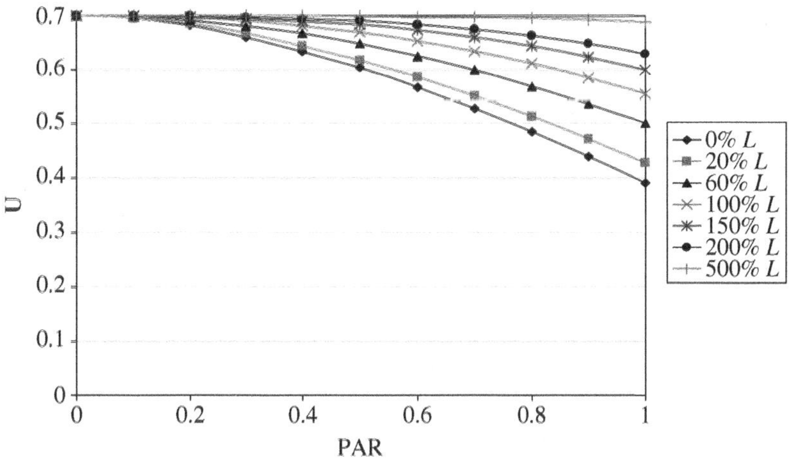

Figure 3.15 The effect of PAR and L on resource utilization. Source: Adapted from Farooq et al. (2005b).

of more requests being accepted for execution per unit time and U is observed to increase.

3.3 Grid Scheduling with Deadlines

An algorithm that performs both allocation and scheduling at the same time on a system with multiple processing units (nodes) is presented. The Grid Scheduling with deadlines (GSD) algorithm introduced in Farooq et al. (January 2009) considers a shared resource b with N identical nodes, a set of jobs $\{i, j, \ldots, k\}$ with sets of the earliest start times $\{t_i, t_j, \ldots, t_k\}$, execution times $\{e_{\mathrm{E},ib}, e_{\mathrm{E},jb}, \ldots, e_{\mathrm{E},kb}\}$, deadlines $\{\mathrm{d}_i, \mathrm{d}_j \ldots, \mathrm{d}_k\}$ and sizes $\{s_{ib}, s_{jb}, \ldots, s_{kb}\}$ where $s_{ib} \leq N$ for all i. The problem is to schedule jobs $\{i, j, \ldots, k\}$ on resource b such that each job i starts executing after its earliest start time t_i, finishes before its deadline d_i. Note that throughout its execution of $e_{\mathrm{E},ib}$ units of time, s_{ib} nodes are allocated to job i.

A high-level description of the algorithm is presented in Figure 3.16.

List A contains all the ARs and ODs that are currently active on the system. When a new request (job) arrives the algorithm computes A and inserts the new job in A, which is then ordered by using a specific ordering heuristic such as EDF (step 1). The first job is then removed from A and its scheduled time is determined on each node of the resource. Determine the earliest scheduled time (after t_i) where job i can be allocated at least s_{ib} nodes (step 2). If by scheduling the job at the scheduled time determined in the previous step its deadline d_i cannot be met

Step 1: Whenever a new job arrives, determine the list A of ARs and ODs to be scheduled and order them using a job-ordering heuristic.
Step 2: Remove the first job i from the ordered list A and determine its scheduled time on each node of the resource. Find the earliest scheduled time of the job after t_i where it can be allocated at least its required number of nodes.
Step 3: If by scheduling job i at the scheduled time determined in Step 2, it cannot be completed before its deadline reject the new request, keep the previous schedule and terminate the algorithm. Note that if any previously scheduled job i cannot be completed before its deadline, it is the new request which is rejected (not the job i).
Step 4: Determine the set of nodes that can schedule job i at its scheduled time determined in step 2.
Step 5: If the number of nodes found in Step 4 is greater than the size of the job i, use a node-selection heuristic to choose the nodes to schedule the job on. Schedule i on the selected nodes at the determined scheduled time.
Step 6: If list A is not empty, go to Step 2. Otherwise, accept the new request and update the resource schedule.

Figure 3.16 High-level description of the GSD algorithm. Source: Farooq et al. (January 2009)/with permission of Elsevier.

then reject the newly arrived request; keep the previous schedule and terminate the algorithm (step 3). Otherwise go to step 4 and determine the set of nodes that can be allocated to job *i* at its scheduled time determined in step 2. If the number of nodes found in step 4 is greater than its size s_{ib} use the node-selection heuristic such as first fit (FF) and best fit (BF) and allocate the selected nodes to the job. Schedule job *i* at the scheduled time determined in step 2 on the selected resources (step 5). If list A is not empty, go back to step 2. Otherwise, accept the new request and update the resource schedule.

The EDF algorithm referred to in step 1 schedules jobs in nonincreasing order of deadlines. The FF algorithm used in step 5 allocates the first set of nodes that can be allocated to the job such that its SLA requirements are met, whereas the BF algorithm selects those nodes that minimize the size of the resulting fragments in the resource schedule (Farooq et al. January 2009).

3.3.1 Performance Analysis

The performance of the GSD algorithm achieved with different values of *L* is presented in Figure 3.17. Details of the system and workload parameters used in this simulation-based performance analysis are available in Farooq et al. (January 2009)

Figure 3.17 presents the results of using several combinations of job-ordering and node-selection heuristics. The job-ordering heuristics are described first followed by the node-selection heuristics.

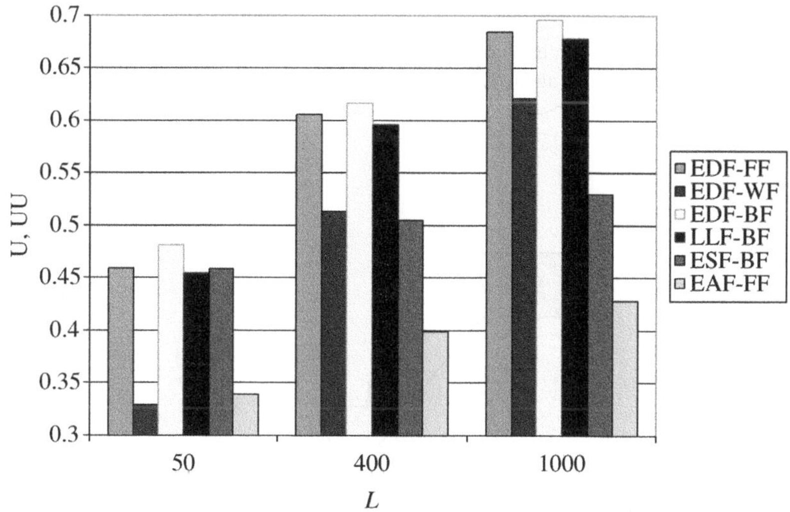

Figure 3.17 Performance analysis of the GSD algorithm. Source: Farooq et al. (January 2009)/with permission of Elsevier.

Job-Ordering Heuristics:

Earliest Deadline First (EDF): orders jobs in a nondecreasing order of their deadlines.

Earliest Start Time First (ESF): orders jobs in a nondecreasing order of their earliest start times.

Least Laxity First (LLF): orders jobs in a nondecreasing order of their laxities.

Node Ordering Heuristics:

First Fit (FF): selects the first node that will be able to accommodate the job.

Best Fit (BF): selects the node that will result in the smallest size fragment in the schedule.

Worst Fit (WF): selects the node that will result in the largest size fragment in the schedule.

The performance metrics displayed in the figure are Utilization (U) and Useful Utilization (UU). U is the proportion of time all the nodes on the cluster/system are busy executing jobs. These jobs may belong to the class of successfully completed jobs or as aborted jobs. UU is the proportion of time all the nodes in the system are busy executing jobs that are successfully completed on the system. For the use case captured in Figure 3.17 U=UU.

As L increases, the flexibility in scheduling improves and U/UU is therefore observed to increase with L. Both the node-selection and the job-ordering heuristics have a significant effect on performance that is observed to change with the

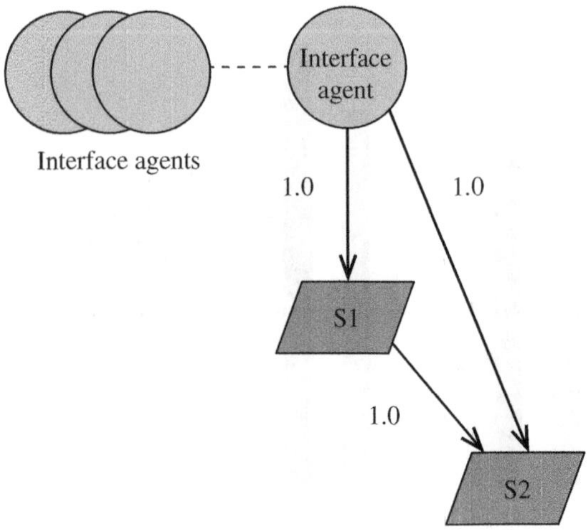

Figure 3.18 Typical interactions on an example client–server system. Source: Graham et al. (1999)/Elsevier.

combination of the two heuristics used. The EDF–BF combination produces the highest utilization for all the values of L presented in Figure 3.17.

3.4 Scheduling on Client–Server Systems

In client–server systems, one or more clients can request service from a common pool of servers each of which can provide a different service. Examples of such servers include file servers, database servers, and print servers. Irrespective of the type of service provided, a client–server system behaves in the following way. The client sends a request to a server through a network. The server performs the desired operation and sends the results of the operation (if any) back to the client.

A typical interaction between a client and server is presented in Figure 3.18 (Graham and Majumdar 1999).

Clients (users) send requests to a server via an *interface agent* that is represented by a circle in the figure. Each Server is represented by a parallelogram. In many situations, a chain of servers is involved in providing a specific service. As shown in the example in Figure 3.18, the request is served by an active server (S1) that executes the required code and invokes another lower-level pure server (S2). The pure server executes and sends the response back to the active server. After getting the result it was waiting for the active server (optionally), executes for some more time, and sends its response request back to the client. An example of an active server is a file system manager that requests service from the disk device (referred to as a pure server). Note that the active server, such as, S1 behaves both like a client and a server. It is a server for the user (client) and a client for the pure server such as S2. A pure server is busy only when it executes, whereas an active

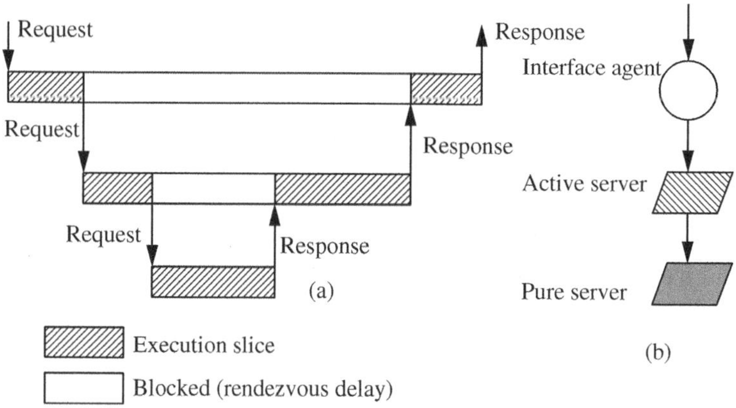

Figure 3.19 (a) Request response diagram. (b) A chain of requests. Source: Graham et al. (1999)/Elsevier.

server may be busy in two ways: when it is executing and when it is waiting for a response from the lower-level server.

The intercommunication between clients and servers is often synchronous that corresponds to remote procedure calls (RPC) and sockets. With such a synchronous communication mechanism, the client is blocked until the response from the lower-level server becomes available. This delay (referred to as a blocked [or rendezvous] delay in Figure 3.19) is shown by the nonshaded rectangles in Figure 3.19 in which shaded rectangles correspond to time spent in the execution of the respective processes.

Asynchronous interactions between clients and server processes are possible; but in this chapter, we will focus on systems characterized by synchronous communication between clients and servers that give rise to a number of important scheduling issues that will be addressed in this chapter.

Scheduling policies described in Graham and Majumdar (1999) each of which uses one or more characteristics of the requests that have arrived on the systems are presented next. The characteristics used are:

Arrival Time: is the time at which the request arrives at a given server

Request Service Time: is the CPU time used to serve the request

Path length: of a request at a given server is the number of servers that are already blocked waiting for a response from this server. For example, in Figure 3.19, the path length of the request that arrives at the active server is zero while that arrives the pure server is 1.

The scheduling policies that may be used at servers are described next.

First Come First Served (FCFS): This policy schedules a request using its arrival time. That is, requests are served in the same order in which they arrived.

Shortest Request First (SRF): Uses *a priori* knowledge of request service time. Given a set of requests the request with the smallest service time is selected for execution on the server.

Longest Path First (LPF): When choosing the next request for service at a given server, the request with the longest path length is selected for execution.

Longest Path First/ Shortest Request First (LPF/SRF): This policy uses a combination of two request characteristics: path length as well as request service time. It works in the same way as LPF, but when two requests have the same (longest) path lengths, it breaks the tie in favor of the request with the smaller service time.

Both preemptive and non-preemptive versions of the scheduling policies are possible. Each thread at a server that executes a request is associated with a priority depending on the scheduling policy in use.

When executing a request, the thread inherits the priority of the request it is serving. For the non-preemptive policies, when a thread completes the service of a request the highest priority thread at the server is executed next. Preemptive scheduling policies are applicable to multithreaded systems only. For a preemptive policy, the execution of a thread is preempted when a higher-priority thread becomes ready to run on the server.

3.4.1 Performance Analysis

A simulation-based performance analysis is provided in Graham and Majumdar (1999) and is described here to compare the performance of the scheduling policies described earlier. The two-server systems shown in Figure 3.18 may correspond to a system that comprises a compute server (S1) and a file server (S2). Each interface agent corresponds to a specific client. The label on the edge connecting the nodes in Figure 3.18 corresponds to the average number of times the node at the head of an edge invokes the other node at the tail of the edge. In the use case modeled in the figure, each request on an average invokes both servers S1 and S2 one time. For every invocation, S1 invokes S2 one time on average. A system subjected to an open stream of client requests is provided by the simulation model. Each client request arrives via its own interface agent, and an aggregate stream of requests is presented to S1. Each server is single-threaded. The aggregated arrival rate for the client request stream is varied. A detailed discussion of the systems and workload parameters is not important for the following discussion but is available from Graham and Majumdar (1999).

The average response time achieved with the non-preemptive versions of four scheduling policies for various arrival rates is presented in Figure 3.20. The response time of a request is computed by subtracting the arrival time of the request from its completion time. As the aggregated arrival rate increases, system contention increases, and the average response time achieved with all the

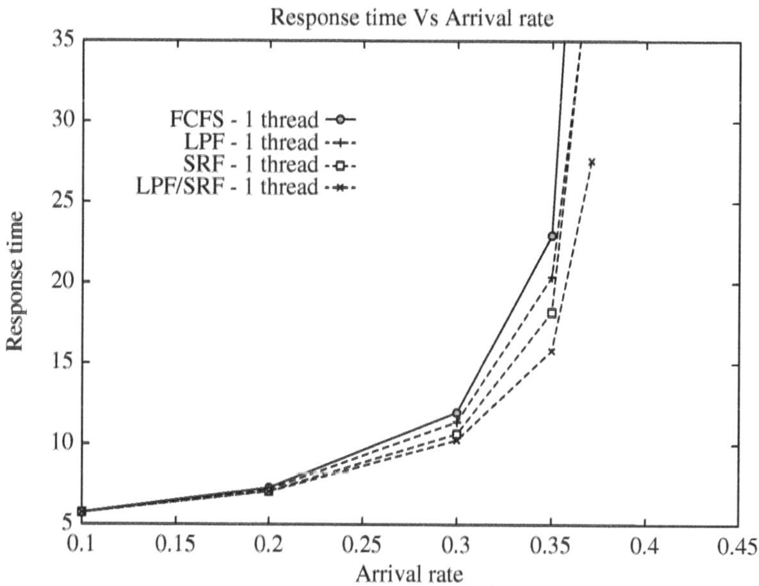

Figure 3.20 Performance of scheduling policies for single-threaded servers. Source: Graham et al. (1999)/Elsevier.

scheduling policies increases. The difference in the performance achieved by the scheduling policies is observed to increase with an increase in the arrival rate.

At higher arrival rates, LPF/SRF performs the best followed by SRF, LPF, and FCFS. Using path length and service time seems to produce an improvement in performance. Using only the request execution time provides higher performance in comparison to the policy that uses only the request path length. Using both of these characteristics in conjunction with one another produces the best performance.

3.4.2 Software Bottlenecks

System performance on a system is often limited by the saturation of a hardware device, such as the CPU and disk. The device with the largest demand saturates and reaches the utilization of 100% when the workload intensity reaches a certain threshold. On a client–server system, such a limit on scalability can be caused by a software server rather than a hardware device. Consider a client–server system that is using synchronous communication among servers. When an active server blocks waiting for a response from a lower-level server that it has invoked, it remains blocked until the response is returned by the lower-level server. Consider, for example, the active server S1 in Figure 3.18. When it invokes S2, it remains

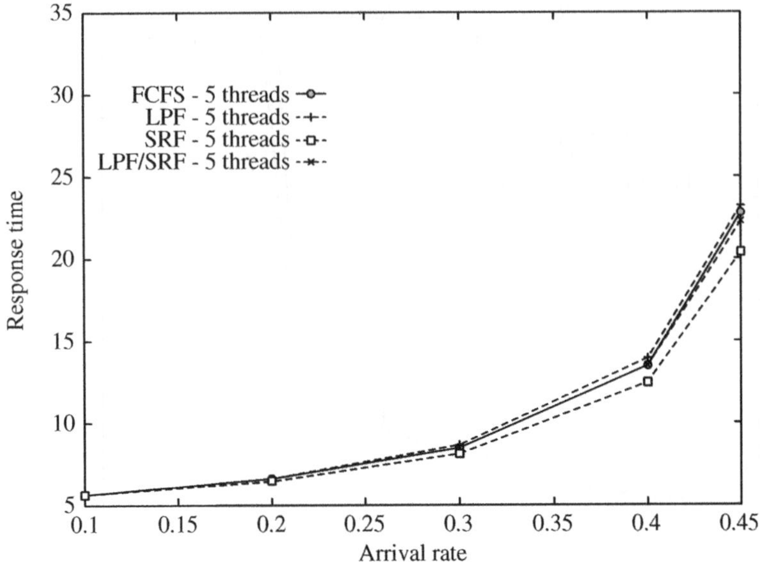

Figure 3.21 Performance of scheduling policies for multithreaded servers. Source: Graham et al. (1999)/Elsevier.

blocked while S2 is serving its request, and the CPU on which it is running remains unused. As a result, after a certain value of the arrival rate is reached S1 remains utilized 100% of the time either executing or waiting for service from S2. The system is then saturated, and the bottleneck is the active server S1. When a system has a software bottleneck (Neilson et al. 1995), using a higher-speed CPU does not alleviate the problem. Multithreading the server can be used for solving the software bottleneck problem.

The performance of the four scheduling policies as obtained by simulation experiments is presented in Figure 3.21. Multithreading the servers seems to give rise to an improvement in performance in comparison to the system using single-threaded servers. It is interesting to note that SRF produces the best performance, whereas LPF/SRF displayed the highest performance in Figure 3.20 exhibits an inferior performance in comparison to SRF.

3.5 Summary

This chapter discusses techniques for allocation of processors to processes and algorithms for process scheduling. Processor allocation techniques are discussed first. Both processor allocation algorithms and their impact on system performance are described. A number of well-known algorithms include Hu's algorithm that produces an optimal allocation for a certain class of precedence graphs as well as heuristic techniques such as LPTF and Multifit are discussed. A set of guidelines for the devising of heuristic allocation algorithms is introduced.

Scheduling when requests are to be satisfied on a best-effort basis is described. This is followed by a discussion of scheduling on grids and clouds subjected to a workload comprising AR requests. The SSS algorithm for systems with a single resource subjected to ARs and the GSD algorithm for systems characterized by multiple resources on which ARs are allocated and scheduled are described. Detailed performance analysis of the techniques and the insights resulting from the analysis are discussed.

The last part of the chapter focuses on client–server systems. Various techniques for server scheduling on a multilevel system comprising both active and pure servers are discussed. System performance and how it can be adversely affected by software bottlenecks are described. Techniques such as multithreading to mitigate the impact of software bottlenecks on performance are discussed.

Exercises

1 In this question, you will be concerned with two parallel applications each of which can comprise four or more processes and runs on a system comprising multiple processors. The average parallelism A1 of Application 1 is <u>smaller</u> than A2, the average parallelism of Application 2. Determine whether the following statement is *true or false*:

If S1 (N) and S2 (N) are the speedups achieved by Application 1 and Application 2, respectively, for N processors and any given scheduling discipline, then:

$$S1\ (N) < S2\ (N) \text{ for any given } N.$$

If the statement is true, prove that it holds true for any number of processors and any scheduling discipline used with <u>both</u> Application 1 and Application 2. If the statement is false provide an example (including the precedence graph and process execution times for Application 1 and Application 2) for which the statement does not hold. Include sufficient documentation to explain your claim.

2 As captured in Graham's anomaly, parallel systems can display anomalous behavior: increasing the number of processors allocated to an application can be detrimental to system performance.

Consider the use of the *Processor Sharing* scheduling discipline for running an application comprising multiple processes on N processors. With such a scheduling, is it possible for the speedup achieved for the application to decrease when the number of processors allocated to the application is increased from N?

If your answer is NO, prove that such a decrease in speedup can never occur on the system (for any application and any value of N).

If your answer is YES, provide an example in which such a decrease in speedup is observed.

3 In *work-conserving* scheduling, a processor is never kept idle if there are one or more ready processes on the system.

Can a nonwork-conserving schedule be the optimal schedule that produces the highest speedup for some concurrent application running on a non-preemptive environment on some given number of processors?

If your answer is NO, then prove that a nonwork-conserving schedule can never be the optimal non-preemptive schedule for any concurrent application.

If your answer is YES, show with the help of an example that such a nonwork-conserving schedule can actually be the optimal non-preemptive schedule for a given application and system.

(Assume that the system you are concerned with contains four processors and the applications that run on the system contain nine or more processes).

4 This question focuses on the application of Hu's algorithm. Consider the precedence graph shown in Figure 3.1. The execution time for all the tasks is two seconds except for tasks P4 and P5 the execution times of which are four seconds. The parallel application corresponding to the precedence graph is to be executed on four processors. Hu's algorithm is to be used in the allocation of processors to tasks.

Determine the completion time for the parallel application. Show all your work including each step of the iterative algorithm.

5 Consider the precedence graph and process executions for an application shown in Figure 3.1 and Table 3.1, respectively. Determine the application completion time when the LPT algorithm is used for processor allocation and when the number of processors is equal to three.

6 Repeat question 5 for a system in which a first come first served process allocation policy is used. With such a policy the processes are executed in order of the times at which they became ready to run. Ties are broken in favor any of the candidate processes.

7 Refer to Figure 3.14 that shows the variation in R_{OD} as PAR is changed. Explain the nonmonotonic behavior captured in the graph. Why does the nonmonotonicity appear for only certain values of L?

8 Refer to Figure 3.20. Explain why the average response time achieved by the scheduling policies is observed to increase with an increase in the arrival rate?

References

E.G. Coffman, M.R. Garey, D.S. Johnson, "An Application of Bin-Packing to Multiprocessor Scheduling", *SIAM Journal on Computing*, Vol. 7, Issue 1, 1978, pp. 1–17, doi: https://doi.org/10.1137/0207001, ISSN 0097-5397.

U. Farooq, S. Majumdar, E.W. Parsons, "Dynamic Scheduling of Lightpaths in Lambda Grids," *Proc. the 2nd IEEE International Workshop on Networks for Grid Applications (GRIDNETS'05)*, Boston, USA, October 2005a, pp. 540–549.

U. Farooq, S. Majumdar, E.W. Parsons, "Impact of Laxity on Scheduling with Advance Reservations in Grids," *Proc. the 13th IEEE International Symposium on Modeling, Analysis and Simulation of Computer and Telecommunication Systems (MASCOTS)*, Atlanta, USA, September 2005b.

U. Farooq, S. Majumdar, E.W. Parsons, "Achieving Efficiency, Quality of Service and Robustness in Multi-Organizational Grids", *Journal of Systems and Software (Special Issue on Software Performance)*, Vol. 82, Issue 1, January 2009, pp. 23–38.

I. Foster, C. Kesselman, C. Lee, R. Lindell, K. Nahrstedt, A. Roy, "A Distributed Resource Management Architecture that Supports Advance Reservations and Co-Allocation," *Proc. of the 7th International Workshop on Quality of Service*, London, UK, May 1999.

R.L. Graham, "Bounds for Certain Multiprocessing Anomalies", *Bell Systems Technical Journal*, November 1966, pp. 1563–1581.

R.L. Graham, "Bounds on Multiprocessors Timing Anomalies", *SIAM Journal on Applied Mathematics*, Vol. 17, March 1969, pp. 416-429.

W.C. Graham, S. Majumdar, "Performance of Scheduling Strategies for Client-Server Systems", *Journal of Parallel & Distributed Computing*, Vol. 58, Issue 3, September 1999, pp. 389–424.

T.C. Hu, "Parallel Sequencing and Assembly Line Problems", *Operations Research*, Vol. 9, Issue 6, 1961, pp. 841–848.

J.E. Neilson, C.M. Woodside, D.C. Petriu, S. Majumdar, "Software Bottlenecking in Client-Server Systems and Rendezvous Networks", *IEEE Transactions on Software Engineering*, Vol. 21, Issue 9, September 1995, pp. 776–782.

A. Silberschatz, G. Gagne, P.B. Galvin, "*Operating System Concepts*", Wiley, NJ, USA, 2014.

4

Resource Management on Systems Subjected to Uncertainties Associated with Workload and System Parameters

4.1 Introduction

The resource management techniques discussed in the previous chapter often assume that exact *a priori* knowledge about applications and system resources is available. Real systems, however, are often subjected to uncertainties associated with both application workloads and resource characteristics. This chapter focuses on two types of such uncertainties that are important in the context of resource management. The first concerns uncertainties associated with task execution times whereas the second with the knowledge of the local resource management policies used by the various resources in a large distributed system.

Estimates of job execution times are often used in resource management (Smith 2004). Resource management for advance reservation (AR) requests, require to use the execution time estimates provided by the user. Such requests are susceptible to errors in estimation. Unless specified otherwise, the terms task and job will be used interchangeably in this chapter. It is hard for a user to provide an accurate estimate of job/task execution times. Moreover, the run time of a job is often dependent on the input data and can thus vary from one run to another. The resource schedule generated by the resource allocation and scheduling algorithms is thus subject to error and may lead to deadline misses and deterioration in system performance. The resource management techniques need to be robust so that they can effectively handle such uncertainties in the task execution time estimates provided by the user. A discussion of such error-handling techniques is provided in Sections 4.2 and 4.3. Handling of overestimation of task execution time is discussed in Section 4.2 whereas the handling of an underestimation of task execution times is discussed in Section 4.3.

Another source of uncertainty occurs in large distributed systems such as cloud data centers that often comprise hundreds or thousands of resources. Such a data center deploys multiple servers, with each server hosting multiple virtual machines (VMs). The local operating systems used on the resources can be

Resource Management on Distributed Systems: Principles and Techniques,
First Edition. Shikharesh Majumdar.

different from one another. The resource management for jobs often depends on the allocation and scheduling algorithms (such as the ones described in Chapter 3 focusing on resource allocation and scheduling) as well as the scheduling algorithm used by the local operating system running on the resource. Handling such variability in local resource management strategies used by the different local operating systems complicates resource management even further and is discussed in Section 4.4. Section 4.5 focuses on techniques for mitigating the detrimental effects of not using knowledge of local resource management policies on performance. The performance improvement resulting from such techniques is described in Section 4.6. Systems that use a special type of resource request called soft advance reservation requests (SARs) are discussed in Section 4.7. How SARs can be effectively utilized by resource management techniques to reduce the negative impact of errors in task execution time estimates on performance is presented. A summary of the key issues discussed in the chapter is included in Section 4.8.

4.2 Handling Errors Associated with User Estimates of Job Execution Times

User estimates of job execution time are often error-prone. The inaccuracies in user estimates of job execution times and their impact on the performance of scheduling policies are reported in Tang et al. (2010) that also discusses techniques for adjusting such inaccuracies and demonstrates the effectiveness of these techniques using real workload traces. The research described in Tsafrir et al. (2007) also points to the existence of errors in user-provided job time estimates and advocates the use of system-generated estimates based on past history in job scheduling. Previous work also shows that user estimates submitted to a supercomputing center were often overestimated. Errors associated with user estimates in job execution times in IBM SP2 traces are described in Mua and Feitelson (2001). This paper discusses the relationship between the performance of a backfilling scheduling strategy and the accuracy of user estimates of job execution times. Underestimation of job execution times can also occur on systems. The impact of overestimation of job execution times on performance and techniques for mitigating their adverse effects are discussed in Section 4.2.1. The impact of underestimation of job execution times and techniques for handling such errors are presented in Section 4.3.

4.2.1 Overestimation of Job Execution Times

Techniques for estimating job execution time are discussed in the literature. Both statistical analyses of applications that have already been run earlier as well as analytical modeling have been used for the prediction of job execution times. But they have resulted in only rough estimates of application run times (Smith 2004). Overestimation of execution times of jobs can lead to deterioration in system utilization. When an AR is accepted the resource manager reserves the resource for the job from a time that is greater that its earliest start time for a duration equal to the estimated job execution time provided by the user. If a job leaves earlier than expected then the resource remains unutilized during remaining time for which the resource was reserved. Note that such a gap in the resource schedule can result not only from the user estimating the execution time incorrectly but also from situations in which an abnormal job termination occurs due to a run time error. In addition to lowering the resource utilization, overestimation of job run times can also result in lowering the revenue for the service provided that might have rejected other requests that could have been fitted into this idle time interval produced on the resource schedule.

A schedule exceptions manager (SEM) and prescheduling engine (PE) are proposed in Farooq et al. (2009) to mitigate the detrimental impact of overestimation of job execution times and abnormal termination of jobs. The first performs resource management during the run time of the job, whereas the second achieves its goals before the jobs start running on the system. These techniques are discussed in Sections 4.2.1.1 and 4.2.1.2, respectively.

4.2.1.1 Schedule Exceptions Manager

When a request arrives a schedulability analysis is performed by the resource manager and the resource on which the job will be executed and the scheduled time at which the job will start execution is determined. As the job runs on the system, SEM monitors the resource schedule. If a job completes unexpectedly before its scheduled exit time, the job the scheduler is invoked and it reschedules another job and tries to utilize the spare capacity that has become available on the resource. Such a rescheduling operation can prepone the execution of previously scheduled jobs as long as their SLA requirements are not violated. SEM is observed to give rise to an improvement in system performance, which is discussed in a later section.

4.2.1.2 Prescheduling Engine

Although SEM can lead to the filling up of idle time slots created in the resource schedule when a job leaves the system unexpectedly, if the execution times of a large number of jobs are overestimated, the system might be forced to reject a number of jobs upon their respective arrivals. This can lead to a decrease in the useful utilization of resources as well as a lowering of the service provider's revenue. The prescheduling engine's objective is to tackle this problem by either overbooking a resource or artificially reducing the estimates of job execution times provided by

Step 1: When a job arrives, artificially reduce the user-estimated job execution time.

Step 2: Perform schedulability analysis.

Step 3: If the results of the analysis show that the proportion of jobs that will miss their deadlines is less than a predetermined threshold, then accept the job. Otherwise, reject the job.

Figure 4.1 High-level description of the overbooking mechanism.

the user before performing resource allocation and scheduling and the respective schedulability analysis.

PE uses an algorithm called OB that can give rise to overbooking a resource and, thus, occasional miss of deadlines. However, using such an OB mechanism can decrease the number of rejections of jobs resulting from overestimation of execution times for jobs. Consider, for example, with a single resource. A job J1 with a user-estimated execution time of 120 minutes was allocated a resource from 9:00 a.m. to 11:00 a.m. When Job J2 with the earliest start time of 10:00 a.m., estimated execution time of 30 minutes and a deadline of 10:45 a.m. arrives, it will be rejected by the system because the resource was already booked for J1 from 9:00 a.m. to 11:00 a.m. If the job execution time for J1 was overestimated and it finishes at 10:00 a.m., the resource will become free from 10:00 a.m. to 11:00 a.m. If J1's execution time were not overestimated Job J2 could have been allocated a 30-minute period during 10:00–11:00 a.m. But since the premature completion time of J1 was not known J2 was rejected leading to a potential decrease in resource utilization and revenue. A high-level description of the algorithm is presented in Figure 4.1.

Additional algorithms for tackling the overestimation of job execution times are possible. For example, the past history of the user can be recorded both in terms of the expected job execution times specified by the user and the actual run time that occurred when the job was run on the system. The ratio between these estimated run times and actual job run times can be saved. Whenever a new job is submitted by this user, this ratio can be used to adjust the estimate provided. For example, if a user has been overestimating job execution times by $x\%$ on average, the submitted job execution time can be reduced by $x\%$ before taking up the job for schedulability analysis (Phuong et al. 2014).

4.3 Underestimation of Job Execution Times

Users can underestimate job execution times as well. The resource manager performs allocation and scheduling using the earliest start time, execution time, and deadline for a job specified by the user. The scheduled time of the job on the resource at which the job will start executing is determined and the job is expected to end at the time that is equal to the sum of its scheduled time and the estimated job execution time provided by the user. The resource is booked for this time interval and can be allocated to another job afterward. If job execution time is underestimated the job will not be completed by its expected end time at which another job may have been scheduled to run. One of the solutions is to abort the job with an underestimated execution time and give the resource to the job that is scheduled to run next. This will give rise to the wasting of CPU cycles that have been used by the job and decrease the useful utilization of the resource. *The useful utilization (UU)* of a resource is defined as:

UU = CPU time used by the resource for executing jobs that are completed successfully/total observation period.

Thus, UU is the fraction of time used by the resource in running only the jobs that were completed on the system, while the utilization of the resource (U) is simply the fraction of time the resource had been used executing both completed and aborted jobs.

A related metric is *Work Aborted* (W_A) which is the total CPU time used on the resource for executing jobs that were ultimately aborted.

Thus, if T is the total observation interval, then:

$$\mathrm{UU} = \left[(T^{*}U) - W_{\mathrm{A}}\right] / T.$$

The SEM technique discussed in the previous section and deployed in the resource manager can be used for combating the adverse effects of underestimation of job execution times. Two strategies, called feasibility policy (FP) and Penalize Underestimation (PU) Policy, are proposed in Farooq et al. (2009) for handling such situations. Both of these are discussed.

Feasibility Policy: With this policy, SEM tries to give a small additional quantum of time ξ to the job if such a decision does not violate the SLA requirements of other jobs. This gives the job a chance to finish its execution. If such an allocation of time that does not violate the SLA requirements of other jobs is not possible the jobs is aborted. If the job does not complete even after executing for the additional quantum of time SEM can give additional quanta of time to the job. Note that the value of ξ is to be chosen carefully: if it is too small multiple quanta each associated with a system overhead may need to be allocated thus increasing the overhead of the resource manager. An excessively large value of ξ on the other hand can reduce the acceptability of newly arrived jobs because of the resource being booked for additional time for the job that is characterized by an underestimated execution time. Thus, a value of ξ that achieves an effective compromise between these two factors needs to be used.

Penalize Underestimation (PU) Policy: With this policy, the resource manager simply aborts the job with an underestimated job execution time. The rationale

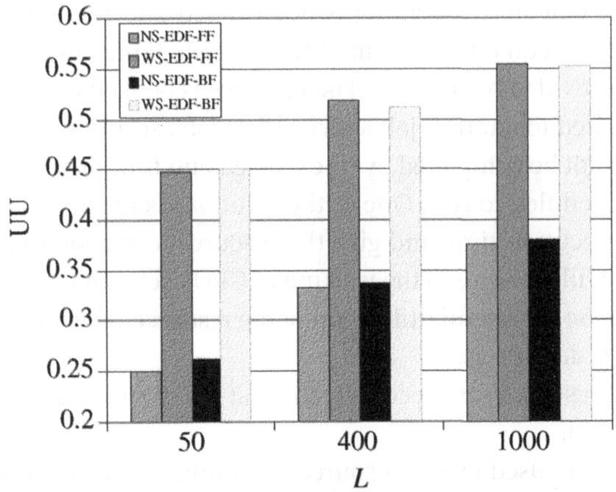

Figure 4.2 The impact of using SEM on useful utilization. Source: Farooq et al. (2009)/Elsevier.

is that since the user is generating an incorrect estimate and not the system is at fault the system should not incur any further overhead in trying to accommodate the user.

4.3.1 Performance Analysis

Analysis of traces collected on IBM SP2 machines and described in Farooq et al. (2009) points to the existence of large errors in the estimation of job execution times. As discussed earlier, overestimation of job execution times by users can lead to unnecessary rejection of other ARs. Figure 4.2 presents a simulation-based analysis of the performance of SEM. The details of the simulation model are available in Farooq et al. (2009) and are not important in the context of the current discussion. The effectiveness of using SEM is captured in Figure 4.2 that presents the useful utilization, UU, for various values of laxity, *L*. As described in Chapter 3, *L* is a measure of slackness, and a higher *L* indicates more slack time available for jobs thus making it easier for the resource manager to meet the deadlines of jobs. The performance of four cases is analyzed, each of which is described by the respective labels. The prefix NS (No SEM) implies that SEM was not used, whereas the prefix WS (With SEM) implies that SEM was used for error handling. The results correspond to the Grid Scheduling with Deadline, GSD algorithm described in Chapter 3 that is characterized by two heuristics: the ordering heuristic and the node selection heuristic. Earliest deadline first (EDF) corresponds to the ordering heuristic, whereas first fit (FF) and best fit (BF) correspond to the node selection heuristic. Each bar in Figure 4.2 represents a specific combination of the ordering heuristic and the node selection heuristic. For example, EDF–FF corresponds to the combination of EDF and FF.

We will consider only the impact of SEM and will thus focus on the NS and the WS cases for each specific heuristic combination. For any given combination,

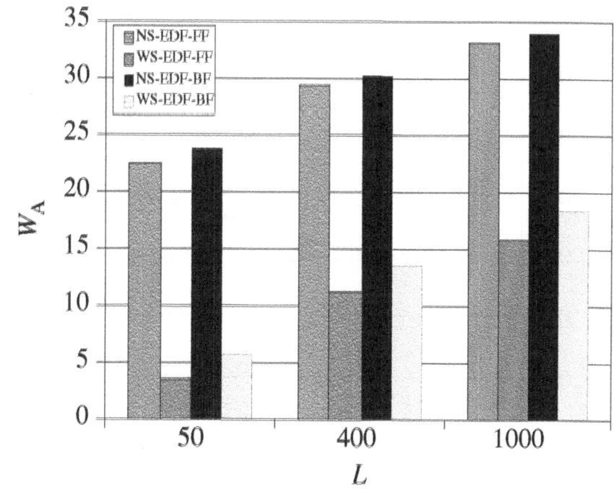

Figure 4.3 The impact of using SEM on work aborted. Source: Farooq et al. (2009)/Elsevier.

EDF–FF or EDF–BF, for example, WS gives rise to a much higher useful utilization in comparison to the NS case. The large improvement in performance shows the utility of using SEM.

The proportion of work aborted (W_A) is observed to reduce significantly when SEM is used. As displayed in Figure 4.3, irrespective of the heuristic combination used the WS cases give rise to a much smaller W_A in comparison to that achieved for the NS case.

The effectiveness of using both PE and SEM is captured in Figure 4.4 in dealing with errors in user-estimated job execution time. Once again, the simulation-based performance evaluation is performed on a system using the GSD algorithm introduced in Chapter 3. Recall that GSD uses two heuristics. In the results presented in the figure, EDF is used as the ordering heuristic and BF is used as the node selection heuristic. Three cases, NE, SEM, and OB are considered. NE corresponds to the ideal case in which users have accurately specified the job execution time, and there is no error associated with user estimates. SEM corresponds to the case in which there is an error associated with user estimates and only SEM is used, whereas OB corresponds to the situation in which there is an error associated with user estimates, and both SEM and PE (deploying OB) are used. The PU policy is used and if a job exceeds its specified execution time it is aborted.

Figure 4.4 shows that the SEM case that does not include PE produces a significantly smaller UU in comparison to the highest achievable performance exhibited by NE. OB that combines both SEM and PE displays a higher useful utilization, UU, in comparison to the situation in which only the SEM is used. This is especially true for the result observed for higher values of γ, the predetermined threshold used by OB. As γ increases, more jobs are accepted by GSD, and its useful utilization increases. The line for OB in Figure 4.4 seems to approach the upper bound

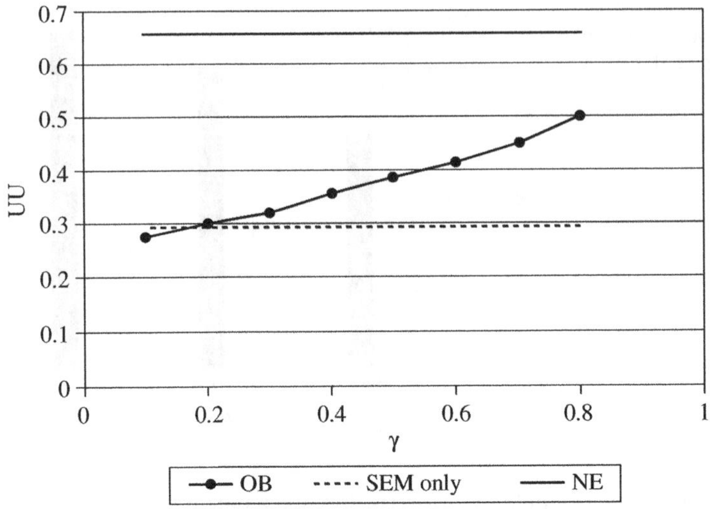

Figure 4.4 The utility of using SEM and OB. Source: Farooq et al. (2009)/Elsevier.

on performance displayed by the ideal case, NE, that achieves the highest possible value of UU.

This improvement in UU displayed in the OB case is, however, accompanied by an increase in R_{AR}, the average response time associated with the AR requests (see Figure 4.5). R_{AR} is the mean response time for ARs, which is the average difference between the time at which an AR completes and its earliest start time. The mean response time is observed to increase with an increase in γ. As noted earlier, a higher γ increases the number of accepted jobs. This, in turn leads to an increase in UU (see Figure 4.4). Higher resource utilization increases contention at the resource, which increases the response time. Note that such an increase in R_{AR} for high threshold values can increase the number of jobs that are pushed beyond their deadlines before completion.

4.4 Handling Uncertainties Associated with the Local Scheduling Policy

The resource management for clouds discussed in the book determines the resource on which a job is to be executed. Scheduling the job to run at a specific time is performed by the operating system deployed on the resource. Scheduling the job thus requires the knowledge of the local resource scheduling policy to ensure that the deadline for the job is going to be met.

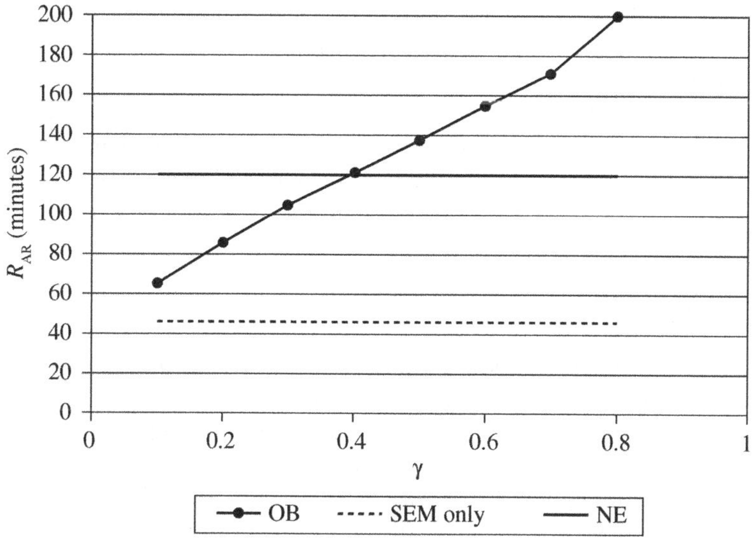

Figure 4.5 The impact of using SEM and OB on the mean response time for AR. Source: Farooq et al. (2009)/Elsevier.

Having *a priori* knowledge of the local scheduling policy at each resource in a large distributed system such as a cloud data center that comprises thousands of resources is complicated due to a number of factors that include the following.

- The exact configuration of the cloud and the details of its constituent servers may not be accurately known at the time the system is deployed. Moreover, the servers used in a cloud are likely to change multiple times during the lifetime of the system.
- Resource Heterogeneity: A cloud data center typically comprises thousands of resources running under multiple different operating systems. Many different types of local scheduling policies for individual resources are potentially in use at the data center.
- Dynamic Nature of Resource Pool: Since resources may be added or removed from the data center at different points in time, it may not always be possible to know the local scheduling policy used at a resource ahead of time at which the resource is added to the resource pool.
- Even if the local scheduling policy is known, it may be time consuming to simulate the operation of the policy. It is also expensive to implement a simulator for all the policies in use in the distributed system.

Performing matchmaking without *detailed a priori* knowledge of the scheduling policy used at a resource can thus mitigate the problems discussed earlier.

This section describes techniques for making such matchmaking decisions without using *a priori* knowledge of the local scheduling policy used by a resource.

Matchmaking techniques that do not use exact *a priori* knowledge of the local scheduling policy of all resources are referred to as *matchmaking in the dark,* that is, discussed later in the chapter. Before proceeding with this discussion, the *Any Schedulability* (AS) *Criterion* used in the matchmaking techniques discussed in the later sections is described in the following section.

4.5 Any Schedulability Criterion

Exact *a priori* knowledge of local scheduling policies at resources may not be available at each resource in a large dynamic system such as a cloud. The AS Criterion is useful in such a situation while deciding on whether or not to accept an incoming AR request. The AS criterion outlines a set of inequalities based on the characteristics of a set of AR requests. Given a set of ARs if the AS criterion is satisfied all the requests in the set are guaranteed to meet their deadlines specified by the user irrespective of the local scheduling policy used at the resource as long as the scheduling policy is a *work-conserving* scheduling policy (Majumdar May 2009). For such a scheduling policy a resource is never kept idle when there is at least one ready process or task available on the system.

Before the AS criterion is described, the terminologies used in the inequalities are introduced.

S_i: Earliest start time for request i

D_i: Deadline associated with request i

E_i: Service time for request i

L_i: Laxity associated with request i

Note: $L_i = D_i - S_i - E_i$

Theorem 4.1 presented in [Majumdar 2009] presents the set of inequalities that characterize the AS criterion.

Theorem 4.1 *A set of requests (i = 1 … N), each of which is characterized by an earliest start time, execution time, and deadline, is any-schedulable if the following condition is satisfied for each request i:*

$$L_i \geq \Sigma_{j \neq i}\ X_1 \cdot min\ \{Ej, (Dj\text{–}Si)\} + X_2 \cdot Ej,$$

$$\begin{aligned}
\textit{where, } X_1 &= 1 \textit{ if} (\text{Sj} < \text{Si, Dj} < \text{Di, Sj} < \text{Di, Si} < \text{Dj}),\\
&\quad \textit{OR if } (\text{Sj} \geq \text{Si, Dj} < \text{Di, Sj} < \text{Di, Si} < \text{Dj}),\\
&\quad \textit{OR } \text{if} (Sj \geq Si,\ Dj \geq Di,\ Sj < Di,\ Si < Dj\ );\\
X_1 &= 0 \textit{ otherwise;}\\
X_2 &= 1 \qquad \textit{if} (\text{Sj} < \text{Si, Dj} \geq \text{Di, Sj} < \text{Di, Si} < \text{Dj});\\
X_2 &= 0 \textit{ otherwise}
\end{aligned}$$

The detailed proof of the theorem is available in Majumdar (May 2009). An explanation of the theorem is provided here. The theorem says that in order for a set of N ARs to be any schedulable, the laxity of each request i ($i = 1 \dots N$) must be greater than or equal to the sum of all the worst-case delays produced by the other requests in the set.

Note that an OD request can be treated as a special case of AR by setting its earliest start time to be equal to the time of arrival of the request and its deadline to be arbitrarily large (infinity).

4.5.1 Application of AS Criterion: An Example

An example of the application of AS criterion in determining the acceptability of ARs is presented with the help of simulation in Majumdar (May 2009). Arrivals of ARs on a resource are simulated. The details of the simulation and workload model are not important in the context of this discussion but are available in the paper. As soon as a request arrives, a set of inequalities comprising the characteristics of the ARs already accepted on the system and the newly arriving AR is constructed. This set of inequalities (one for each AR in the set) is similar to the set described in the theorem. If all the inequalities in the AS criterion are satisfied, the newly arriving request is accepted. Note that the satisfaction of the AS criterion guarantees that all requests will be able to meet their deadlines irrespective of the scheduling policy used at the resource as long as the policy is work conserving.

A number of simulation experiments for a system using the EDF as the resource scheduling policy were performed. EDF, discussed in Chapter 3, is a priority-based scheduling policy that gives a higher priority to a task with a shorter deadline. It is work conserving because it never keeps the resource idle if there is a ready task on the system. The system is investigated for two resource management policies. In the first, the AS-based criterion is used. This is referred to as Anysched + EDF in Figure 4.6 because the AS criterion is used for deciding whether or not to accept the arriving job without using the knowledge of the underlying scheduling policy (EDF) that is used at the resource. The other policy uses the knowledge of EDF in deciding whether or not to accept the newly arriving job. This is referred to as EDF + EDF in the figure. Upon arrival of the request, an EDF algorithm is simulated. The priority of the job is determined. If the simulator indicates that all the jobs in the set of jobs that include the jobs that are already present on the systems and the newly arrived job will meet their SLAs, the job is accepted.

System performance is computed by using a performance metric called blocking ratio, shown as BR in Figure 4.6. The BR for a given request arrival rate is computed as the ratio of the ARs rejected and the total number of ARs simulated. The performance of the two policies is presented in the figure. Figure 4.6a corresponds to the case where the percentage laxity (PL) is 100%, and Figure 4.6b corresponds to a PL

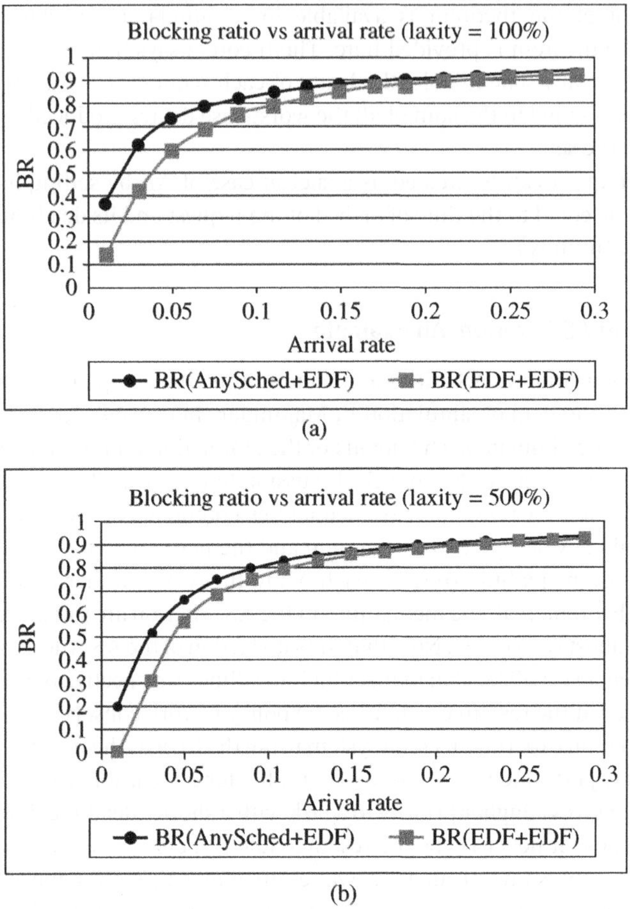

Figure 4.6 Impact of arrival rate: (a) PL = 100%; (b) PL = 500%.

value of 500%. PL is related to the execution time of a job as:

$$\text{Laxity of a job} = \text{Job execution time} * \text{PL}/100.$$

In both Figure 4.6a and b, BR is observed to increase with arrival rate. This is because as the arrival rate increases, system contention increases, and it becomes more difficult to accept a job. As a result, BR increases. For both values of PL, 100% (Figure 4.6a) and 500% (Figure 4.6b) the difference between the two policies becomes smaller as the arrival rate increases. For any given arrival rate, this difference is smaller for PL = 500%. It is interesting to note that the AS criterion-based resource management policy that does not use the knowledge of the local scheduling policy (EDF) performs comparably with the other resource management policy that uses previous *a priori* knowledge of the local scheduling policy at the resource at higher arrival rates. This demonstrates the usefulness of the AS criterion that is used for matchmaking algorithms for clouds, which are discussed in the following section.

4.6 Matchmaking in the Dark: AS Criterion-Based Matchmaking

Matchmaking in the dark refers to matchmaking on systems without having exact *a priori* knowledge of the local scheduling policies used at all resources on the system. The AS criterion is used in such cases. AS criterion-based matchmaking uses Theorem 4.1. When a request arrives on the system and if the AS criterion is satisfied by any of the resources, the request is accepted. The request is rejected otherwise. If there is one resource on the system, then the request is accepted if and only if AS criteria are satisfied on the resource. It is rejected otherwise. If there are multiple resources that satisfy the AS criterion, then a fitness criterion is sued for selecting which of the resources will be used. The fitness criteria used include the following:

First Fit (FF): Select the first resource from the list of resources on which the AS criterion is satisfied for the jobs allocated to the resource.

Next Fit (NF): is similar to FF except that the search in the list of resources that satisfy the AS criterion starts from the resource next to the one that was selected in the previous allocation.

Best Fit (BF): Select the resource from the list of resources on which the AS criterion is satisfied, and the slack for accommodating future requests is minimized.

4.6.1 Hybrid Matchmaking

Techniques that use the AS criterion along with additional scheduling strategies are referred to as hybrid techniques and will form the subject of discussion for this subsection. As discussed earlier, cloud data centers typically comprise a large number of resources. The resources are classified into two categories: *transparent* and *opaque* (see Figure 4.7). The transparent resources are resources for which the behavior of the local scheduling policy is known in advance, whereas the scheduling policy used for the opaque resources is unknown. A hybrid matchmaking algorithm for such a system is described in Melendez and Majumdar (2010).

As discussed in this paper, the presence of opaque resources in the resource pool occurs due to one or more of the following reasons:

- including newly acquired resources for which the scheduling simulators are not available currently and may take a considerable time to develop;

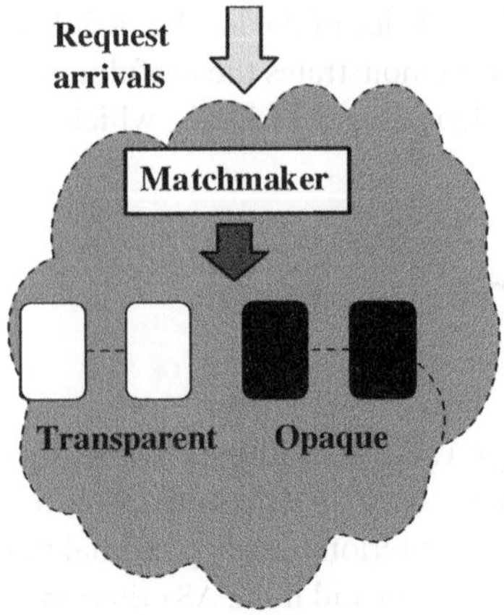

Figure 4.7 Two classes of resources in a cloud. Source: Melendez and Majumdar (2010)/with permission of IEEE.

- including a set of resources acquired temporarily from another organization or department in an enterprise for dynamically accommodating spikes in computing resource demand (as described in Kim et al. 2009).

A hybrid matchmaking strategy that uses one matchmaking algorithm for the transparent and resources and the AS-based criterion for the opaque resources is discussed. If the knowledge of the local scheduling policy is not available, the opaque resources cannot be used in a system that does not use AS criterion-based scheduling. Hybrid matchmaking mitigates the problem by using a matchmaking policy that uses knowledge of the local scheduling policies on transparent resources and an AS criterion-based matchmaking on the opaque resources. Two types of hybrid matchmaking strategies are proposed in Melendez and Majumdar (2010). They are discussed in Sections 4.6.1.1 and 4.6.1.2. The performance of these strategies is discussed in Section 4.6.2.

4.6.1.1 Independent

The transparent resources are managed by a *transparent broker* and the opaque resources by an *opaque broker* (see Figure 4.8). The transparent broker uses the knowledge of local scheduling policies used at the resources. When a request arrives it uses a simulator for the local scheduling policy used at a resource to determine whether the resource which may have been booked for other ARs at different points in time will be able to meet the SLA requirements of the new request. If the resource can meet the SLA requirements the request is accepted; otherwise, the request is rejected. If multiple resources can accept the request, the fitness criterion is used to determine which of these resources will be allocated to the new request. The opaque broker, on the other hand, is not aware of the local scheduling policy used at the opaque resources and uses AS-based matchmaking

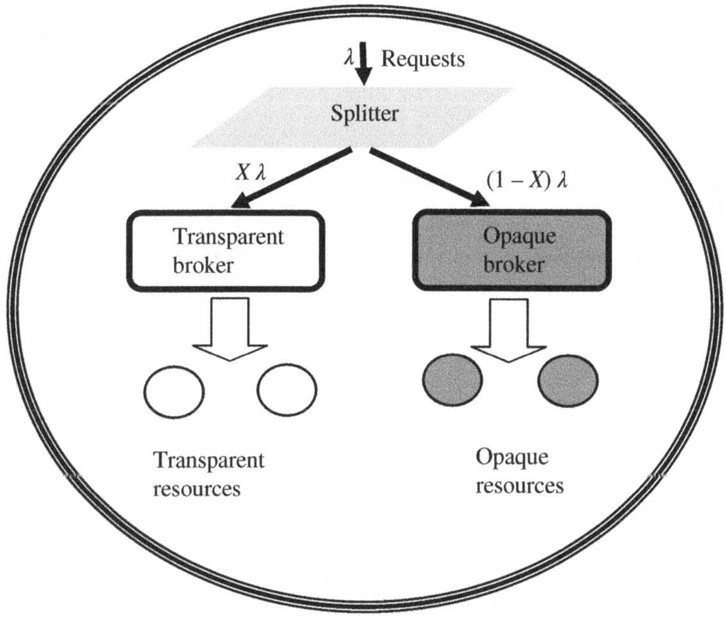

Figure 4.8 The independent strategy.

and determines whether or not the request is to be accepted and which of the opaque resources is to be used if the request is accepted.

The incoming request stream is split by a splitter into two request streams: one sent to the transparent and the other to the opaque broker. Upon arrival of a request, the transparent broker is selected with a probability of X and the opaque broker with a probability of $1 - X$. Thus, if the overall request arrival rate is λ, the arrival rate for the stream directed at the transparent resource is $X.\lambda$ and that for the stream directed at the opaque resources is $(1 - X).\lambda$. One possible way to choose the values of X is to ensure that the load is balanced at the transparent and opaque set of resources.

4.6.1.2 Combined

In the independent matchmaking technique, a request is sent either to the transparent or to the opaque broker. This may give rise to a situation in which a request rejected by the transparent or the opaque broker is rejected right away without checking whether the request could be accepted by the other broker. Thus if, for example, a request sent to the transparent broker cannot be accepted for any transparent resource, the request will get rejected even if there is room on one of the opaque resources to meet its SLA.

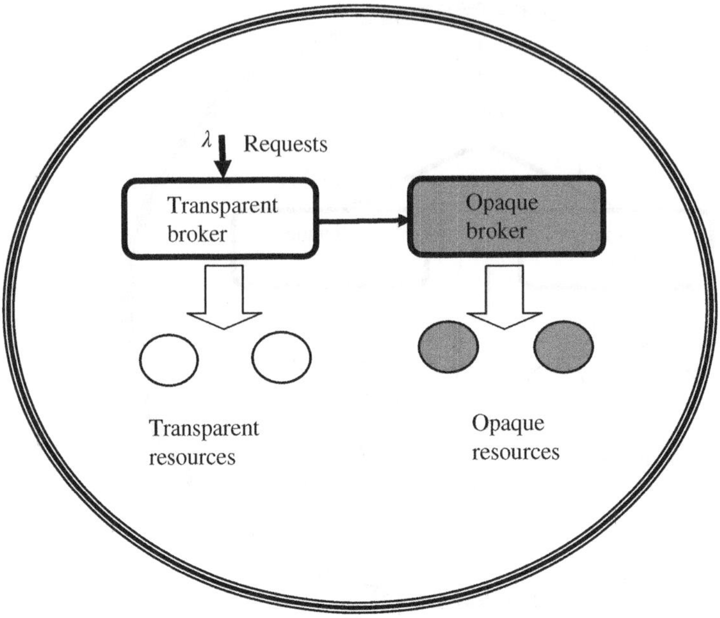

Figure 4.9 The combined strategy.

This problem is alleviated in the combined strategy (see Figure 4.9). For this strategy, the request is sent to only the transparent broker and if the request cannot be accepted by this broker it is sent to the opaque broker. If the SLA for the request can be met by one of the resources managed by the opaque broker, the request is accepted; it is rejected otherwise.

4.6.2 Performance Comparison of Independent and Combined Matchmaking

A simulation-based performance analysis of the two techniques is presented in Melendez and Majumdar (2010) that includes a discussion of the simulation model and the synthetic workload used. The performance metrics used for the performance comparison are described first.

Blocking Ratio (B): is the ratio of the number of jobs rejected and the total number of jobs submitted. This is the same as BR described in Section 4.5.1.

Rate of Accepted Jobs (A): is the number of jobs that are accepted per unit time. It is related to the blocking ratio B:

$$A = (1 - B).\lambda$$

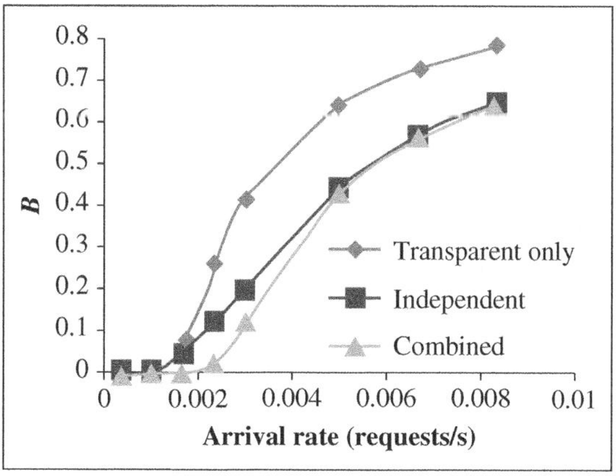

Figure 4.10 The comparison of hybrid matchmaking techniques: blocking ratio. Source: Melendez and Majumdar (2010)/with permission of IEEE.

Revenue Rate (R): Every job accepted on the system earns revenue for the service provider. *R* is the revenue earned per unit of time by the service provider. It is dependent on *A* as well as the revenue model used. In the analysis presented here, a linear relationship between CPU time and the revenue earned is assumed. Thus, if the service provider charges cunits for every CPU second, then the total revenue earned is obtained by measuring the busy time of the resources and multiplying by cunits. Note that cunits is simply a measure of currency used such as dollar, euro, or cent.

The simulation experiments described in Melendez and Majumdar (2010) are for a system comprising 10 resources (i.e. $N = 10$). The proportion of transparent resources PT is held at 0.5. Figure 4.10 shows the variation in *B* as a function of the arrival rate. EDF is the local scheduling policy used at the resources. The Transparent Only line in the figure corresponds to the case in which EDF-based matchmaking policy is used. Concomitantly only the transparent resources could be used. The figure showed a large decrease in *B* achieved with the two hybrid matchmaking techniques: independent and combined especially at higher arrival rates.

The corresponding acceptance rates achieved in the three cases are captured in Figure 4.11. Lower *B* achieved by the hybrid matchmaking techniques in comparison to the Transparent Only leads to a higher acceptance rate. The acceptance rate achieved by the simpler independent technique is comparable to that of combined technique especially at low and high arrival rates. At medium values of arrival rates, combined seems to lead to a slightly higher *A* in comparison to independent.

The higher values of *A* translate to a significantly higher revenue rate, *R*, for the service provider (see Figure 4.12). At the highest arrival rate displayed in the figure, a 60% improvement in revenue rate over independent only is demonstrated by the two hybrid matchmaking techniques.

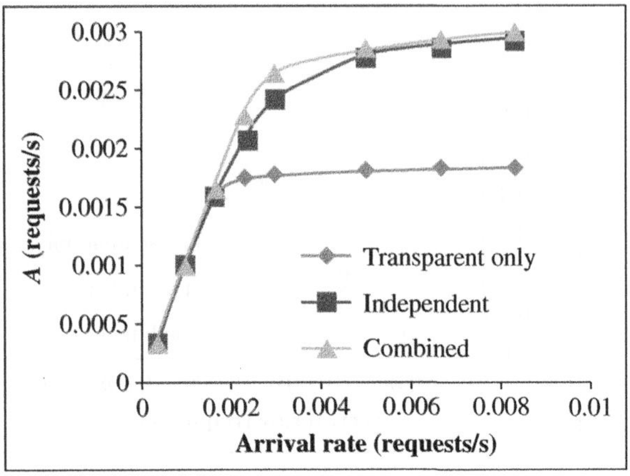

Figure 4.11 The comparison of hybrid matchmaking techniques: acceptance rate. Source: Melendez and Majumdar (2010)/with permission of IEEE.

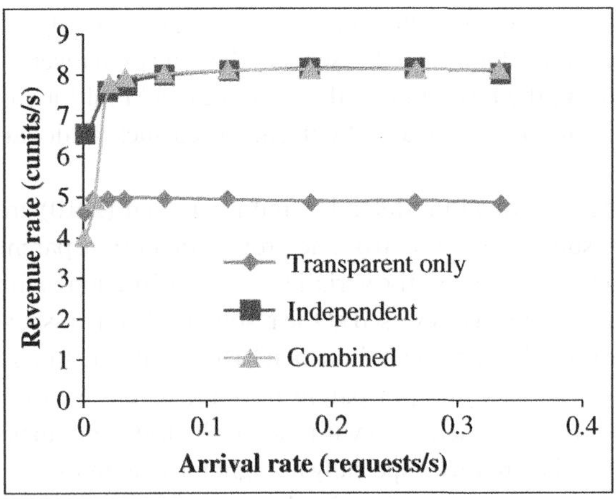

Figure 4.12 The comparison of hybrid matchmaking techniques: revenue rate. Source: Melendez and Majumdar (2010)/with permission of IEEE.

4.7 Soft Advance Reservation Requests

Advance reservation requests require that the deadline specified by the user be met at all times. A Soft Advance Reservation Request relaxes the condition and allows a given proportion of ARs to miss their deadlines. Such an SAR approach-based matchmaking is described in Phuong et al. (2014). A description of the technique is presented first. This is followed by a discussion of its performance that is presented in Section 4.7.2.

This SAR-based technique is used for handling errors associated with user-estimated execution times discussed earlier. The technique has strategies for dealing with both overestimated and underestimated execution times. It keeps track of the recent history of the requests submitted by the user and of the difference between the user-estimated run times specified as part of the SLA by the user and the actual run times for the jobs that occurred in the system. Upon arrival, the technique can alter the user-specified execution time by replacing it with a system-generated execution time by using the past history of job execution.

Although this mechanism can be combined with any matchmaking technique described earlier, analyses of systems that use AS-based matchmaking are discussed in this chapter. In this way, the SAR-based technique is robust both in terms of its ability to handle errors in user-specified job execution times as well as being able to refrain from using the knowledge of local scheduling policy used at the resources.

An important input for the algorithm is a grade of service parameter called P_{miss} specified by the user. The algorithm ensures that the proportion of requests that miss their deadlines is always less than this parameter. It keeps track of R_{miss} the proportion of requests that have missed their deadlines so far. Whenever a request arrives, it decides to replace the user-specified job execution time with a system-generated estimate of execution time (discussed later in this section) if:

$$R_{miss} \leq (P_{miss} - \delta),$$

where δ, referred to as the margin is a threshold parameter used to ensure that the proportion of the deadline never exceeds P_{miss}. If the inequality is not satisfied, the algorithm uses the user-specified run time. The rationale is that if R_{miss} is too close to P_{miss} then it is better to avoid the risk of a new deadline miss because of the error in system-generated estimate of job execution time.

4.7.1 Computation of System-Generated Estimate of Job Execution Time

This technique is aimed at mitigating the detrimental effects of overestimation of job execution times by the user. The computation of the system-generated estimate of job execution time is based on statistics of the last W jobs executed on the system. The information on past history used by this algorithm is the percentage of overestimated execution times for the past W requests (α) to reduce the user-estimated execution times for the incoming jobs. Since requests with underestimated runtimes provided by users lead to job runtimes equal to their user-estimated runtimes, they are not used in determining α that is computed as:

$$\alpha = \left(\sum_{k=1}^{W} O_k\right)/W,$$

O_k is the proportion of overestimated runtimes that is computed as:

$$O_k = [(T_E - T_R)/T_E] * 100,$$

1. Decide whether to use the user-specified estimate or the system-generated estimate of job run time.
2. If the system-generated estimate is to be used, go to step 3. If the user-estimated execution time is to be used, go to step 4.
3. Compute a system-generated estimate of job execution time by using the technique described earlier.
4. Use the AS criterion to determine the list of resources that can satisfy the SLA of the request.
5. If multiple resources can satisfy the criterion, then use the fitness criterion to select the resource on which the job will execute.
6. If the list generated in step 4 is empty reject the request.

Figure 4.13 The SAR-based match-making approach.

where T_{E} is the user-estimated run time of the kth request (job), and T_{R} is its actual run time.

The system-estimated job execution time, T_{M}, is then computed as:

$$T_{\mathrm{M}} = T_{\mathrm{E}} * (1-\alpha/100).$$

The key steps of the matchmaking algorithm are described in Figure 4.13.

As discussed in Section 4.6 multiple fitness criteria such as First Fit, Next Fit, or Best Fit are possible candidates. For the simulation results described later, FF is used.

The handling of jobs that exceed their estimated run time is discussed next. Two algorithms are proposed in Phuong et al. (2014).

Algorithm 1 aborts a job if its execution time so far has exceeded its user-specified execution time. However, if the run time so far is equal to the system-generated estimate but still less than the user-specified estimate, the job is preempted and is resumed at the first available time slot on the resource. If this time slot is not long enough for the job to complete it is allocated another free time slot at the resource. This continues until the job's run time has reached the user-specified job execution time or the job has been completed. Otherwise, the job is ejected.

Algorithm 2 does not abort a job when its runtime has reached the time it has been allocated on the resource. Irrespective of whether it has exceeded the system-generated or user-specified run time estimate it is prompted and resumed during a free time slot in the schedule for the resource. If it still does not complete the process is repeated until the job completes irrespective of whether or not the job misses its deadline.

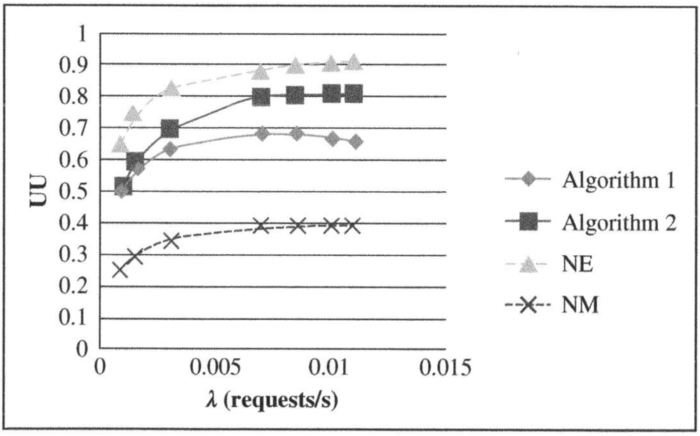

Figure 4.14 Impact of arrival rate on the performance of SAR-based matchmaking. Source: Phuong et al. (2014)/with permission of IEEE.

4.7.2 Performance Evaluation

A simulation-based performance analysis of the SAR-based approach is described. The details of the simulation and workload models are provided in Phuong et al. (2014). In addition to the performance of Algorithms 1 and 2, the performance of two other systems NM in which no modifications to user-specified estimates of job execution times are made and NE that corresponds to the ideal situation in which there is no error associated with the user-specified estimates are included in the graphs.

As shown in Figure 4.14, the useful utilization UU of the system increases with an increase in arrival rate λ. This is because as λ increases more and more jobs are executed on the system and the utilization of the resource increases. Both Algorithms 1 and 2 give rise to a significant improvement in UU in comparison to NM demonstrating the effectiveness of the algorithms in tackling erroneous estimation of job execution times by users. The performance of Algorithm 2 is observed to approach the ideal case (see NE) with an increase in λ. For a given λ, Algorithm 2 gives rise to a higher performance in comparison to Algorithm 1.

Figure 4.15 shows the relative performance of the four systems when the laxity L in the jobs in the workload is varied. As L increases the slack times in jobs increase and more jobs are accepted by the AS criterion-based matchmaking algorithm. Consequently, the useful utilization of the system increases. The higher performance of Algorithms 1 and 2 for all L values in comparison to NM demonstrates the effectiveness of the two algorithms devised to handle errors in user estimates of job execution times.

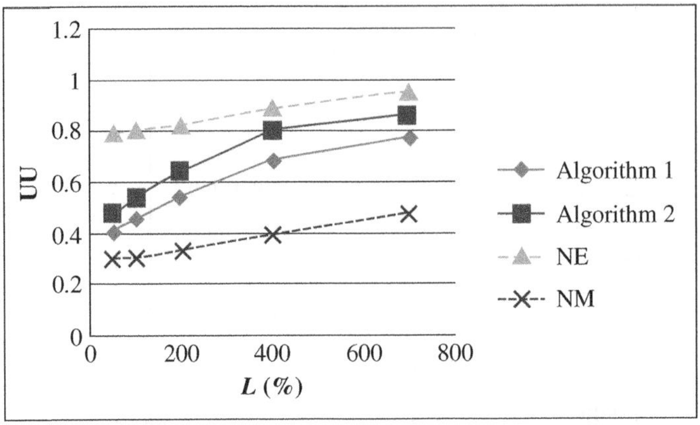

Figure 4.15 Impact of laxity on the performance of SAR-based matchmaking. Source: Phuong et al. (2014)/with permission of IEEE.

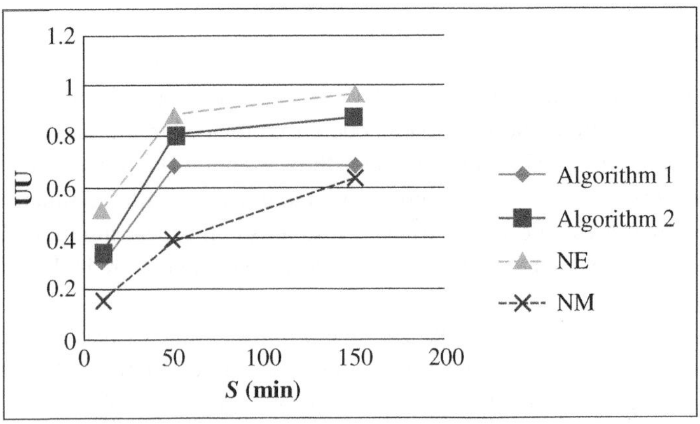

Figure 4.16 Impact of service time on the performance of SAR-based matchmaking. Source: Phuong et al. (2014)/with permission of IEEE.

UU of the system is observed to increase with an increase in the mean job execution time S (see Figure 4.16). Once again, for any given S, Algorithms 1 and 2 outperform NM and Algorithm 2 displays a higher performance in comparison to Algorithm 1. This shows the utility of preventing job abortions by Algorithm 2 even if the jobs miss their deadlines.

4.8 Summary

This chapter concerns two types of uncertainties associated with the system: those that are related to workload parameters and those that concern the servers in the system. The first results from the errors associated with estimates of task execution times provided by the user. The impact of both overestimation and underestimation of task execution times that can result in unnecessary rejection of AR requests and job abortions are discussed. Two error-handling techniques that can mitigate such detrimental effects of estimation errors on performance, SEM and PE, are described. An analysis of the performance of these techniques is also provided. Another way to combat the effects of errors in job execution time estimates is using soft ARs that allow a certain proportion of job requests to be rejected. The SAR concept, resource management techniques for systems subjected to requests characterized by SARs, and how they can lead to performance improvement are described.

Another type of uncertainty discussed in the chapter stems from the lack of knowledge of the scheduling policies used by the local operating systems running at the different servers. Two types of system classifications: transparent and opaque are introduced. The local scheduling policies at transparent resources are known to the resource manager, whereas those used at opaque resources are unknown. An AS Criterion that is useful for matchmaking on opaque resources is introduced. Two different matchmaking techniques that utilize the AS criterion are described. A discussion of the performance benefits that can accrue from these techniques is provided.

Exercises

1 Using examples distinguish between utilization and useful utilization of a resource. Discuss the impact of each on system performance.

2 Discuss the role of the prescheduling engine and SEM in controlling the impact of error in user-estimated job execution times. For each discuss if they are useful in the context of both overestimation and underestimation of job execution times.

3 Consider the following set of ARs:

AR #	Earliest start time	Execution time	Deadline
1	10:00 a.m.	60 min	1:00 p.m.
2	9:00 a.m.	50 min	10:30 a.m.

Does this set of ARs satisfy the AS Criteria?

4 A company is planning to purchase a resource management system. A certain supplier claims that if its proprietary system is used, then the deadline for the following sets of ARs will be met even when there is a single processor in the system.

AR #	Earliest start time	Execution time	Deadline
1	11:00 a.m.	50 min	1:00 p.m.
2	12:30 p.m.	60 min	2:00 p.m.
3	11:30 p.m.	120 min	2:30 p.m.

The Systems Engineer for the company says that the supplier cannot deliver what it is promising. Explain why.

5 What changes in the characteristics of the AR(s) described in question 5 need to be made so that the Systems Engineer changes their mind. Note that all the details of the resource management techniques are not known to the Systems Engineer.

6 Explain the role of the AS criteria in hybrid matchmaking.

7 Consider Figure 4.12. Discuss the possible rationale behind the performance of Transparent, Independent, and Combined. Why is the improvement in *R* that accrues for the hybrid matchmaking techniques more pronounced at higher arrival rates?

8 Explain with the help of examples the difference between an AR and a Soft AR. Explain why the introduction of Soft ARs can lead to an improvement in performance.

References

U. Farooq, S. Majumdar, E. Parsons, "Achieving Efficiency, Quality of Service and Robustness in Multi-Organizational Grids", *Journal of Systems and Software (Special Issue on Software Performance)*, Vol. 82, Issue 1, January 2009, pp. 23–38.

H. Kim, S. Chaudhari, M. Parashar, C. Marty, "Online Risk Analytics on the Cloud", *Proc. of the IEEE/ACM International Symposium on Cluster Computing and the Grid*, Shanghai, China, 2009.

S. Majumdar, "The "Any-Schedulability" Criterion for Providing QoS Guarantees Through Advance Reservation Requests", *Proc. IEEE/ACM International Symposium on Cluster Cloud and the Grid (International Workshop on Cloud Computing)*, Shanghai, China, May 2009, pp. 490–495.

J. O. Melendez, S. Majumdar, "Matchmaking with limited Knowledge of Resources on Clouds and Grids", *Proc. International Symposium on Performance Evaluation of Computer and Telecommunication Systems (SPECTS),* Ottawa, July 2010, pp. 102–110.

A. W. Mua'lem and D. G. Feitelson, "Utilization, Predictability, Workloads, and User Runtime Estimates in Scheduling the IBM SP2 with Backfilling", *Parallel and Distributed Systems, IEEE Transactions on*, Vol. 12, Issue 6, June 2001, pp. 529–543.

H. Phuong, S. Majumdar, M. Zaman, P. Srivastava, N. Goel, "Resource Management Techniques for Handling Uncertainties in User Estimated Job Execution Times", *Proc. 2014 International Symposium on Performance Evaluation of Computer and Telecommunication Systems (SPECTS'14),* Monterey, USA, July 2014, pp. 626–633

W. Smith, "Improving Resource Selection and Scheduling Using Predictions," in: *Grid Resource Management. International Series in Operations Research & Management Science* (eds: J. Nabrzyski, J.M. Schopf, and J. Węglarz), Vol. 64. Springer, Boston, MA, 2004, pp. 237–253.

W. Tang, N. Desai, D. Buettner, Z. Lan, "Analyzing and Adjusting User Runtime Estimates to Improve Job Scheduling on the Blue Gene/P", *Proc. of the 2010 IEEE International Symposium on Parallel & Distributed Processing (IPDPS)*, 2010, pp. 1–11.

D. Tsafrir, Y. Etsion, D. G. Feitelson, "Backfilling Using System-Generated Predictions Rather than User Runtime Estimates", *Parallel and Distributed Systems, IEEE Transactions on*, Vol. 18, Issue 6, 2007, pp. 789–803.

5

Resource Auto-Scaling

5.1 Introduction

Resource provisioning requires the determination of an adequate number of resources appropriate for meeting the desired performance objectives of an institution, such as a research center or a data center in an enterprise. It needs the availability of appropriate tools, skills, and experience to achieve this objective. The problem is further complicated by the variation of workload intensity with time. As a result, if one provisions a number of resources for the worst-case situation, the peak load (see line 2 in Figure 5.1), resources are likely to remain idle during other times when the workload intensity is lower and a smaller number of resources are adequate (see line 1). Provisioning resources for the average and lower workload intensities (line 3) may lead to a poor quality of experience for the users during high load periods. Auto-scaling or changing the system capacity in line with current workload requirements (see line 4) is an important technique deployed in many data centers and is the subject of attention in this chapter. The system capacity can be changed in two different ways that are discussed. Changing the system capacity by changing the capacity of individual resources is referred to as vertical auto-scaling, whereas achieving the change in system capacity by increasing or decreasing the number of resources is referred to as horizontal auto-scaling.

The types of requests that characterize the system workload are introduced in Section 5.2. Three classes of horizontal auto-scaling techniques are discussed in the remaining sections. Reactive auto-scaling techniques change the number of resources based on a change of system state that is often determined by a change of key performance metrics such as resource utilization, mean job turnaround time, and the proportion of incoming requests that could not be accepted because of the inability of the system to meet request deadlines. Reactive auto-scaling techniques and the relationship of their performance with system and workload parameters are discussed in the first part of Section 5.3.

Resource Management on Distributed Systems: Principles and Techniques,
First Edition. Shikharesh Majumdar.

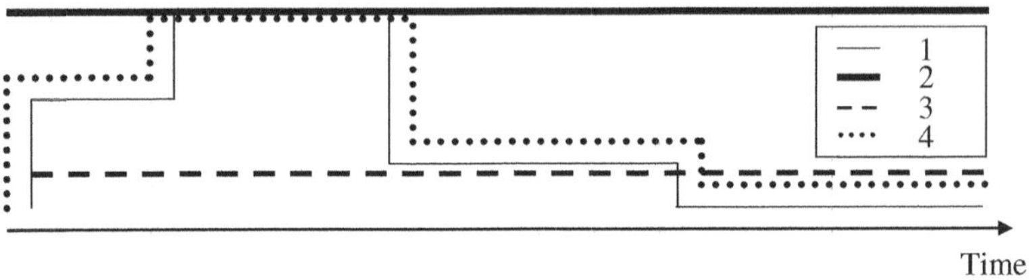

Figure 5.1 Variations in the number of resources with the workload requirements.

While reactive auto-scaling makes a decision on resource management after the change in system state has occurred, proactive auto-scaling predicts the change in the future system state and proactively determines the adequate number of resources that will be required at the respective future point in time. Proactive auto-scaling algorithms and their performance are described in the second part of Section 5.3. A discussion of a combined approach called hybrid auto-scaling is presented in Section 5.4. The highlights of the discussion presented in this chapter are summarized in Section 5.5.

5.1.1 An Example System

Provisioning an adequate number of resources is important in the context of any Information Technology (IT) department in an enterprise. Consider the system presented in Figure 5.2 in which a single client enterprise (SCE) rents resources from an intermediary cloud provider (IP) that creates a virtual private cloud (VPC) for the enterprise by using resources acquired from a public cloud provider (CP). Although the resources are acquired from the public cloud, they are used exclusively by the users in SCE, hence, the term VPC. Note that the role of IP can also be assumed by the IT department in the enterprise, serving clients from various other departments in the institution. The broker in IP provides various control functionalities, including the scaling of resources used by SCE. Further details of the broker are provided later in the chapter in the context of the respective auto-scaling techniques discussed.

As the intensity of the workload produced by SCE increases more resources are often required to meet the client service level agreements (SLA) and/or the grade of service (GoS) requirement for the system. The SLA may include parameters such as a deadline for completion of the jobs that are accepted by the system. To avoid deadline violations, service providers need to occasionally reject user requests if the required SLA cannot be met with existing system resources. Rejecting requests leads to a loss of revenue by the intermediary CP. Moreover, the intermediary CP often needs to limit the number of requests rejected in order to maintain a predetermined level of Grade of Service (GoS) required by SCE. Resources from the CP

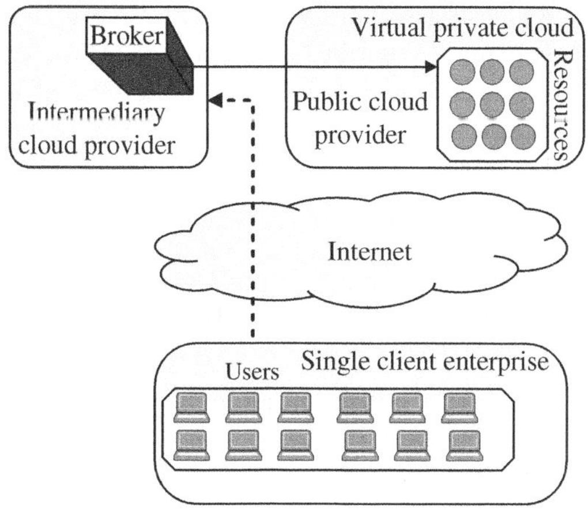

Figure 5.2 The system using auto-scaling. Source: Biswas et al. (2015a)/with permission of IEEE.

thus need to be acquired for increasing the revenue earned by the intermediary CP and/or for meeting such GoS requirements on the system. When the workload intensity is reduced all the existing resources in the private cloud are not required and some of the resources can be returned to the public CP by the intermediary CP for minimizing the cost it incurs for renting the resources from the public CP.

As discussed earlier, there are two types of auto-scaling: *horizontal and vertical*. In horizontal auto-scaling, the number of resources in the VPC is changed in accordance with the current system load. In vertical auto-scaling, the capacity of an existing resource is changed by changing its speed and/or memory capacity. The rest of the chapter focuses on techniques for performing horizontal auto-scaling. Before discussing the auto-scaling techniques, the various characteristics of the requests submitted by the user are discussed.

5.2 Request Characteristics

A request submitted by a user in SCE is associated with an SLA specified by the user. Each SLA is characterized by a number of components that are described next.

- Earliest Start Time (EST) – is the earliest time at which the request can start execution on a resource.
- Service Time (ST) – is the total CPU time required by the request. All resources are assumed to be statistically identical and thus the CPU time required by the request to complete is the same on each resource.
- Deadline (DL) – is the request completion deadline, the time by which the request execution must complete.

Requests associated with such SLA parameters are similar to Advance Reservation requests described in Foster et al. (May 1999). Note that requests that are not associated with completion deadlines are often referred to as on demand (OD) requests. In the following discussion, such requests are considered to be a special case of AR requests associated with a deadline value set to infinity (a very large value).

5.3 Horizontal Auto-Scaling

Three types of horizontal auto-scaling are discussed in the literature, each of which is described next.

Reactive Auto-Scaling: Resources are added or removed from the VPC *after* a predefined change in the workload has occurred.

Proactive Auto-Scaling: Resources are added or removed from the VPC proactively before the actual change in workload has occurred. Such techniques are often based on algorithms that try to predict the future workload and increase/decrease the number of resources in anticipation that certain predicted changes in workload are about to occur.

Hybrid Auto-Scaling: The system combines both the reactive and proactive auto-scaling approaches and comprises two components: a reactive component and a proactive component.

Each type of auto-scaling system is discussed.

5.3.1 Reactive Auto-Scaling

In reactive auto-scaling, a change in the number of resources is made after a change in state has occurred on the system. For example, an additional number of resources may be acquired when the workload intensity or a specific performance metric crosses a predetermined threshold value, and some of the existing resources may be released and returned to the public CP when the workload intensity falls below a given threshold value. Examples of existing reactive auto-scaling systems include Amazon's CloudWatch (Amazon n.d.-a) that uses target metrics for system performance specified by the user to trigger auto-scaling decisions.

Threshold-based reactive auto-scaling is also suggested in Dutreilh et al. (July 2010). This chapter indicates that it is important to carefully tune the thresholds such that oscillations in the number of system resources do not occur. Oscillations may occur if the system keeps on increasing/decreasing the number of resources to satisfy the threshold rules. A cool-down or calm period during which no further scaling operations can be made after a scaling operation has been performed is also described in Dutreilh et al. (July 2010). Auto-scaling rules are usually based only on a single or small number of parameters, such as CPU load, system response time, or the input request rate.

RightScale's auto-scaling algorithm uses the mixing of auto-scaling rules and voting. If the majority of the virtual machines (VMs) decide on scaling up/down, the respective action is taken; no changes in the number of resources are made otherwise (RightScale n.d.).

A broker-based approach to reactive auto-scaling is presented in detail. A performance analysis of the system including the impact of various system and workload parameters is described in a following section. In addition to performing auto-scaling the broker is responsible for allocation of resources to jobs/user requests (matchmaking) and scheduling the requests allocated to the same resource. The different components of the broker that cooperate with one another to achieve its functionality are explained with the help of Figure 5.3.

Request Handler (RH): receives the requests generated by the users and sends them to the decision maker (DM).

Decision Maker (DM): uses the matchmaking and scheduling (MatchMakeSched) module to ensure that the deadline associated with the given request can be met. If the deadline can be met the request is accepted. Otherwise, the reactive auto-scaler (RA) is invoked to determine whether or not a new resource is to be acquired. If a new resource is to be acquired, the request is accepted and allocated to the new resource. Otherwise, the request is rejected.

Matchmaker and Scheduler (MatchMakeSched): applies a matchmaking and scheduling algorithm to determine whether the SLA associated with the newly arriving request can be met with the available resources and the pending requests on the system.

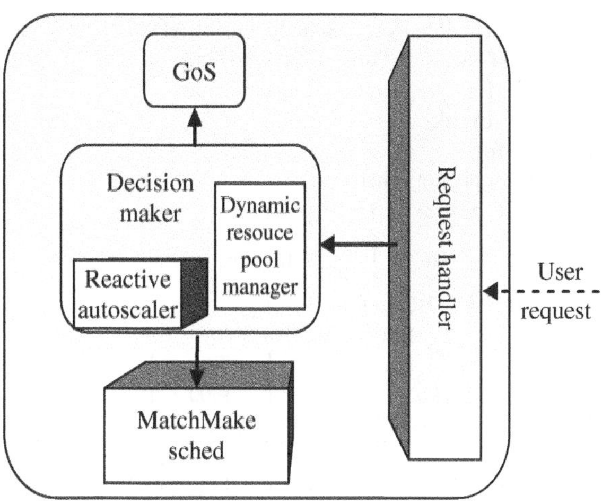

Figure 5.3 The broker architecture for a reactive auto-scaler. Source: Biswas et al. (2015a)/with permission of IEEE.

Reactive Auto-Scaler (RA): is responsible for determining whether or not a new resource is to be acquired or an existing resource is to be released based on the current system state.

Grade of Service (GoS) Module: is used by RA in making its decision on whether or not a new resource is to be acquired.

Dynamic Resource Pool Manager (DRPM): is responsible for managing the dynamically varying set of resources.

5.3.1.1 Reactive Auto-Scaling Algorithm

A reactive auto-scaling algorithm based on the techniques described in Biswas et al. (2015a) is presented in Figure 5.4. The algorithm uses the user cost UC incurred by the user and the broker cost BC incurred by the broker for making auto-scaling decisions. In the price model used in this technique, the user is charged on a per second basis and $c_pvt is the cost *per second* charged by the broker to the user. The broker has to pay the public CP in terms of the number of hours of resource usage and is charged $c_pub *per hour*. Even if the resource is used for part of an hour, the broker is charged for the full hour. A similar pricing model is used for some resource instance types by Amazon, for example (Amazon n.d.-b). Thus, if a request executes for c_i seconds, the user cost is computed as:

$$UC = G \times \lceil c_i \rceil \times \lceil \text{c\_pvt} \rceil),$$

where G is a padding factor that is a tuning parameter with a value greater than 1. It is used to account for potential requests from the user that can execute this newly acquired resource in the future. The cost incurred by the broker for executing this

---

for each resource Res_j, from $j = 1$ to N do
  if Req_i can be scheduled on Res_j
    Schedule Req_i // Schedule using MMS
    if Res_j.stopTime < Req_i.endTime
      Res_j.stopTime ← Req_i.endTime
    break;
end for
if Req_i is not scheduled
  UC ← G × ([c_i] × [c_pvt])
    $BC \leftarrow \left\lceil \frac{C_i}{3600} \right\rceil \times [\text{c\_pub}]$
  if (UC – BC) > 0 or B > B_{spec}
    $\text{start}_j \leftarrow$ (EST of Req_i) - T
    $\text{stop}_j \leftarrow \text{start}_j + \left\lceil \frac{\text{num}Seconds}{3600} \right\rceil$
  acquireResource(start_j,stop_j)

Figure 5.4 The reactive auto-scaling algorithm. Adapted from Biswas et al. (2015a).

request, on the other hand, is computed as:

$$\text{BC} = \left\lceil \frac{c_i}{3600} \right\rceil \times \lceil \text{c\_pub} \rceil.$$

The first term on the right-hand side is the smallest number of hours that is equal to or greater than the request execution time. Note that a new resource j is given a start time start_j and a stop time stop_j where start_j = EST of the new request that will execute on res_j-T and $\text{stop}_j = \text{start}_j + \lceil \text{numSeconds}/3600 \rceil$ and numSeconds is the execution time for the new request.

The algorithm is executed upon the arrival of each request, Req_i (line 2). The currently existing resources on the system are scanned from Res_1 to Res_N. If the SLA for the request can be satisfied on any of the existing resources, it is scheduled to execute on that resource (line 3). If the end time for Req_i exceeds the current stop time for the resource, its end time is extended to the request end time Req_i.endTime (line 5), and the algorithm is terminated. If the request's SLA cannot be satisfied on any of the existing N resources, the user cost, UC, and the broker cost, BC, are computed (lines 9 and 10). If UC exceeds BC implying a profit for the broker a new resource is to be acquired. The auto-scaler attempts to maintain a B that is lower than B_{spec}. Thus, even if UC is lower than BC, a resource is acquired if B becomes higher than B_{spec} (line 11). The start time for Res_j is set to the earliest start time for the request that will execute on the resource first—T (line 12). At this point in time, Req_i is the only request scheduled on Res_j. The subtraction of T seconds from EST of Req_i is performed on line 12 to account for the time required by the CP to start a new resource. Since the newly arrived request is the only request scheduled to be executed on the newly acquired resource, the stop time for the resource is set to the time that is the sum of the resource start time and the smallest duration in hours that is greater than or equal to the execution time of the new request (line 13). The resource is then acquired in line 14.

5.3.2 Performance Analysis of the Reactive Auto-Scaling Algorithm

A simulation-based performance analysis for the algorithm (referred to as System I) was done to compare its performance with that of two other alternate systems, System II and System III (Biswas et al. 2015a). System II is based on a greedy algorithm from the literature (Yu et al. July 2005) that acquires resources when the SLA for the request cannot be satisfied by any of the existing resources without considering whether or not such a resource acquisition will lead to a broker profit. In System III, the user does not go through IP and acquires resources directly from the public CP by paying for the resources on an hourly basis. The performance metrics of interest are User cost, UC and broker profit, BP. Note that the performance of a proof-of-concept prototype running on the

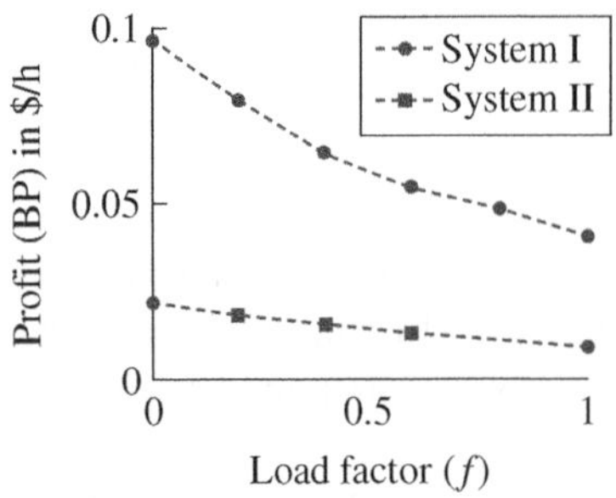

Figure 5.5 Variation of broker profit with f. Source: Biswas et al. (2015a)/with permission of IEEE.

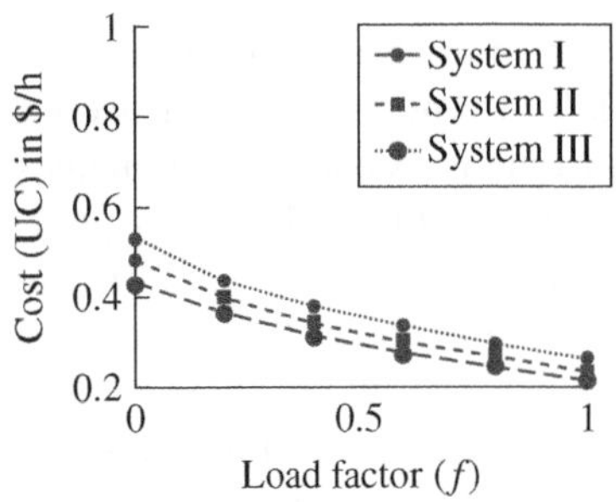

Figure 5.6 Comparison of user costs. Source: Biswas et al. (2015a)/with permission of IEEE.

Amazon EC2 cloud showed close agreements with the simulation results (Biswas et al. 2015a).

The rate of arrival of requests is often observed to vary with the time of the day. The graphs presented in Figures 5.5 and 5.6 are for such a system that is characterized by two arrival rates: a low arrival rate λ_{low} and a high arrival rate λ_{high}. The parameter f is the load factor that is equal to the proportion of time for which the systems are subjected to the lower arrival rate. The specific values of the system and workload parameters, such as the various request and cost characteristics including the distributions used for their generation are available in (Biswas et al. 2015a). Figure 5.5 that compares the two broker-based systems (System I and System II) shows that for both systems BP increases as f decreases. This is because a lower f means that the system is subjected to a higher arrival rate for a larger proportion of time, enabling the broker to earn more profit. Given any value of f, BP for System I is substantially higher than that for System II, demonstrating the effectiveness of the profit-based reactive auto-scaling technique. Figure 5.6 shows the superiority of System I in terms of UC in comparison to that achieved with System II and System III. The lower value of UC achieved with System I for any f in comparison to that achieved with System III, in which the user directly acquires resources from the public CP provides the motivation for the user to use System I, the reactive auto-scaling system utilized by the intermediate cloud service provider.

5.3.3 Proactive Auto-Scaling

A reactive auto-scaler changes resource allocation after a given change in the system is detected in the system state, while a proactive auto-scaler makes a decision about changing resource allocation in advance before the change in system state has occurred and follows through with the decision at a predicted point in time. In proactive auto-scaling, the system attempts to predict a future system state and decide whether or not auto-scaling is to be performed at a given future point in time. Typically, this auto-scaler predicts the future workload and decides whether or not additional resources are to be acquired in the future to handle this change in workload. Since resource acquisition from the public CP takes a finite amount of time a proactive auto-scaler is expected to avoid this delay by acquiring resources in advance by anticipating the future system state. Proactive auto-scaling algorithms may be based on time series analysis, control theory, reinforcement learning, or queuing theory. The workload predictor is often used with a performance model to determine the number of resources required to meet the specified GoS when the system gets subjected to the predicted workload.

Timeseries analysis can be used in devising proactive auto-scaling techniques (Lorido-Botran et al. 2014). The technique uses a time window to store past events that are used to make predictions of future events. The width of the window needs to be chosen carefully because it has an important impact on prediction accuracy. If the window is too short, the prediction accuracy may be poor. On the other hand, a long window increases complexity, leading to a high system overhead. A proactive technique for resource provisioning is described in Moore et al. (April 2013) that uses a real-time performance monitoring framework to make the auto-scaling decisions for a cloud platform. The proactive controller uses three components – a time series forecaster and two incrementally updatable Naive Bayes model. The forecasts made by the time series models are generated by using a Support Vector Regression (SVR) algorithm.

Some of the proactive techniques that use time series analysis to identify patterns in the workload. An example is provided by Gong et al. (October 2010). This paper also describes a broker model with the broker charging the user and broker paying a fee to the CP.

An approach based on autoregressive–moving average (ARMA) model has been used in Bhagavathiperumal and Goyal (2019) for building a proactive workload predictor. This predictor attempts to provision the required resources prior to the increase in resource demand during peak hours. Such an ability of provisioning resources before the actual workload change has occurred is the advantage of proactive auto-scaling techniques that avoid the delay in resource provisioning experienced with reactive auto-scaling techniques that can only start the resource provisioning process after the workload change has occurred on the system.

A proactive technique that uses machine learning is described in more detail in the following sections. The performance analysis of the technique that shows the impact of system and workload parameters on the performance of the auto-scaling system is included.

5.3.3.1 The Proactive Auto-Scaling Algorithm

The broker-based proactive auto-scaling technique described in Biswas et al. (December 2014) is discussed next.

The system architecture for a proactive auto-scaling system is presented in Figure 5.7. Each component of the system is briefly described.

Broker: The broker receives the user request and decides whether or not to accept it. If the SLA of the request can be met with the current resources in the system, the request is accepted; it is rejected otherwise.

MatchMake Sched (MMS): performs matchmaking and scheduling and operates in a fashion similar to that described for the reactive auto-scaler. It selects the resource on which a request is to be executed and determines the start time for the request such that the request's SLA can be met.

Machine Learning Engine (MLE): receives information about request characteristics and predicts the characteristics of requests to arrive in the future using a Machine Learning algorithm that typically uses the past history of request arrivals to predict the characteristics of future requests. Examples of such techniques include linear regression (LR) and SVR.

Decision Maker (DM): uses the output of MLE to obtain the characteristics of the requests that will arrive in the future and uses MMS to simulate the matchmaking and scheduling algorithm for a future point in time. Based on whether or not a broker profit will occur at the said future point in time it makes the decision on whether or not to acquire any additional resources. DM is run periodically to generate the predicted characteristics of the next k requests and make decisions on

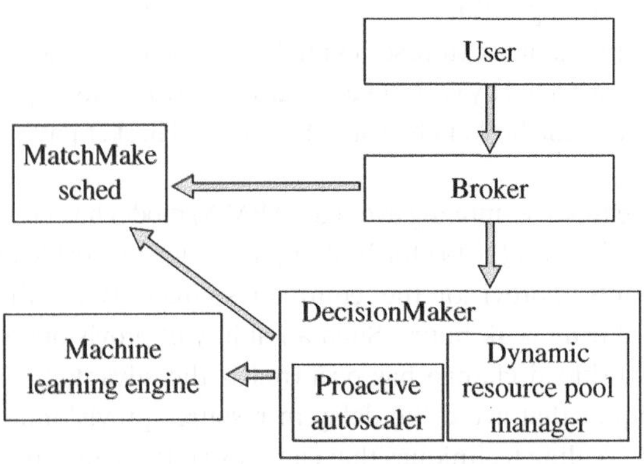

Figure 5.7 The proactive auto-scaling system. Source: Biswas et al. (December 2014)/with permission of IEEE.

resource acquisition and release. As in the case of the reactive auto-scaler *DRPM* is responsible for managing the dynamically varying set of resources.

A high-level approach for the algorithm used by the Proactive Auto-scaler is described next.

The values of the request characteristics that include *Arrival Time* (AT), *Earliest Start Time* (EST), *Service Time* (ST), and *Deadline* (DL) are predicted by MLE. Note that the last three characteristics are the SLA parameters that are expected to be specified by the user. MLE attempts to predict these four characteristics for all the k requests that are expected to arrive on the system next.

DM then simulates the arrival of these k requests. Both the request as well as the resources are simulated. The schedulability of these requests is then analyzed with the help of MMS. For each of the requests it allocates the request on an existing resource if the resource can satisfy the SLA of the request. Otherwise, a new resource is added to the simulated set of resources and the request is scheduled on it.

After completing the scheduling of each of the k requests the profitability analysis is performed for each resource that includes the existing resources as well as resources acquired in the simulation of the k future requests. The key steps of this process are described. For each resource, i:

- *Start time for the ith resource ($start_i$)* is computed by the DM as:

$$\text{start}_i = (\text{EST of the request expected to execute the earliest of the } i\text{th resource}) - T$$

  where T is the additional time required by the public CP for starting a new resource.
- *Stop time for the ith resource ($stop_i$)* is computed as:

$$\text{stop}_i = \text{start}_i + \left\lceil \frac{numSeconds}{3600} \right\rceil,$$

  where numSeconds is the time (in seconds) beyond start_i for which the resource will be used. Note that the ceiling function is used because a resource is to be acquired for an integral number of hours from the public CP.
- *Predicted Broker Cost* (BC_i^p): is the predicted cost (determined by DM) for using the ith resource. It is computed as: $[(stop_i - start_i)] \times \$c\_pub$
- *Predicted service time* ($\mathbf{s}_i$): is the cumulative service time for the predicted q requests scheduled on the ith resource (determined by DM). It is computed as:

$$\sum_{j=1}^{q} st_j^i \text{ in seconds}$$

- *Predicted Total User Cost* $\left(\mathrm{UC}_i^p\right)$: is the cost incurred by the users to execute q predicted requests on the ith resource (determined by DM). It is computed as:

  $$[s_i] \times c_{pvt}$$

- *Predicted Broker Profit* (BP_i^p): is the predicted revenue earned from serving user requests on the ith resource (determined by DM). It is computed as:

  $$\mathrm{BP}_i^p = \left(\mathrm{UC}_i^p - \mathrm{BC}_i^p\right).$$

  (Note that the superscript p is used for predicted values of the respective parameters.)

DM then computes the profitability of each resource. For each resource i, if it is a new resource to be acquired for handling the k requests, DM decides that the resource should be acquired if BP_i^p is greater than zero. If i is an existing resource, DM decides to accept the new request(s) assigned to it if BP_i^p is greater than zero.

The proactive auto-scaling algorithm is executed on a separate thread by the broker. Resources are acquired at the specific start times determined by this algorithm. RH is run on a separate thread. Upon receiving a new request, RH uses MMS, the MatchmakeSched module to perform the allocation and scheduling of the new request. If the SLA for the request can be met, the request is accepted; it is rejected otherwise. Note that the matchmaking and scheduling decisions at this point in time are based on the actual characteristics of the request that arrived in the system.

5.3.4 Performance Analysis of the Proactive Auto-Scaling Algorithm

A simulation-based performance analysis of the proactive auto-scaling algorithm is presented in Biswas et al. (2015b). System I, System II, and System III have already been described in Section 5.3.2. The details of the system and workload parameters used, including the distributions used to generate their values, are available in Biswas et al. (2015b). For the two broker-based systems, System I and System II, as the load factor increases, the broker profit is observed to decrease (see Figure 5.8). As indicated earlier, a higher f indicates a lower proportion of time for which the system is subjected to the higher arrival rate λ_{high}. Thus, a lesser number of requests are processed per unit of time on average for a higher value of f, thus decreasing broker profit. A comparison of user costs for the three systems shows that, for any given value of f, System I leads to the least UC, followed by System II and then System III (see Figure 5.9). This demonstrates the effectiveness of the proactive auto-scaling algorithm for both the Intermediate Service Provider and the user. Not that the superscript a used with the metrics presented on the y-axes denotes that these are for the requests actually arriving on the system.

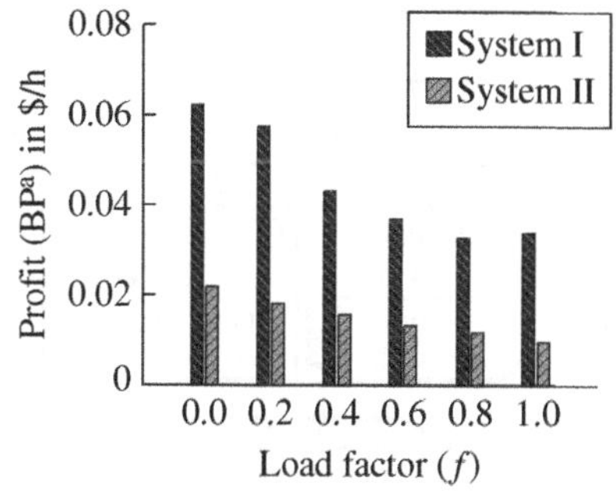

Figure 5.8 Variation in actual broker profit with variation in load factor. Source: Biswas et al. (2015b)/with permission of IEEE.

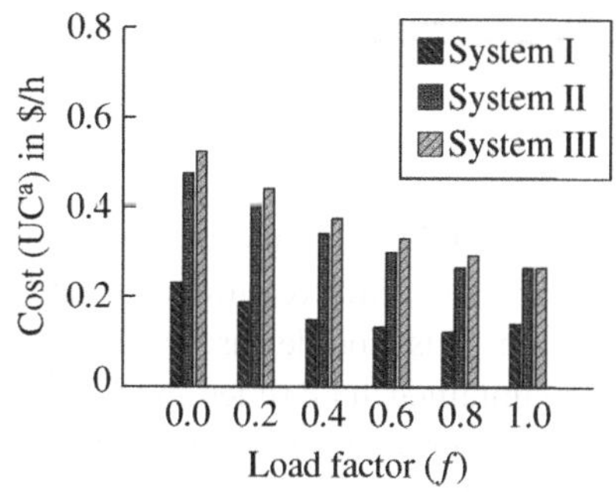

Figure 5.9 Comparison of user costs. Source: Biswas et al. (2015b)/with permission of IEEE.

5.4 Hybrid Auto-Scaling

Reactive and a proactive auto-scaling techniques were discussed in the previous sections. Each of them focuses on increasing broker profit and reducing the user cost. The proactive auto-scaling technique allows the system to make the auto-scaling decision in advance and thus can avoid the delay due to resource provisioning incurred by a reactive auto-scaling technique that performs resource acquisitions right at the time when the detection of a workload change warranties the deployment of additional resources to handle the change. The proactive system, on the contrary, uses predictions of future workload and is thus susceptible to not meeting the desired profit and cost objectives when the actual workload is different from the workload predicted when the prediction algorithm was run. Reactive auto-scaling avoids this situation since it reacts to changes in system workload that have really occurred on the system.

Hybrid auto-scaling combines the advantages of both reactive and proactive auto-scaling.

Figure 5.10 shows the components of the broker for the hybrid auto-scaling system.

The MLE and the DRPM have already been discussed in the context of the reactive and proactive auto-scaling systems. The DM used in the hybrid auto-scaling

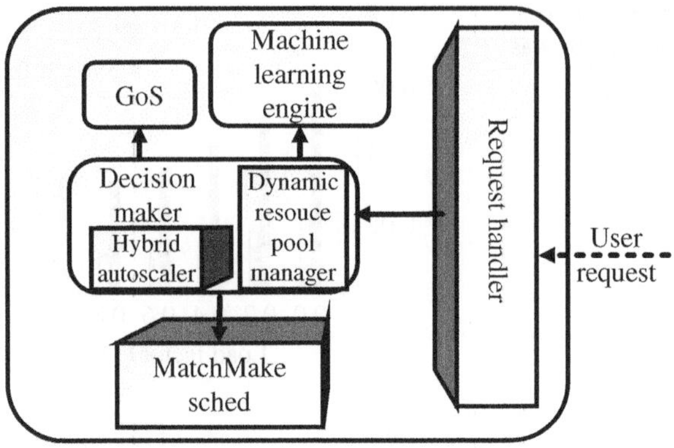

Figure 5.10 A hybrid auto-scaling framework. Source: Biswas et al. (2017)/Springer Nature/CC BY 4.0.

technique uses a hybrid auto-scaling mechanism. The GoS module records the GoS set by SCE and is used to maintain the blocking ratio, B, that is, the ratio of the number of requests that are rejected because of the inability of the system to meet their SLA requirements and the total number of requests arriving on the system. This module ensures that the desired level of GoS, B_{spec}, specified by SCE, is maintained at all times and the blocking ratio B remains lower than B_{spec}.

The hybrid auto-scaler runs two threads, one for reactive auto-scaling and one for proactive auto-scaling. The reactive auto-scaler runs for every request arrival. It decides to acquire a new resource if:

(i) No currently available resource on the system satisfies the request's SLA

AND

The acquisition of the resource will lead to a profit for the broker

OR

(ii) Rejecting the request will lead to a B that is higher than the specified GoS (B_{spec})

The proactive auto-scaler is run periodically, and as discussed earlier, it predicts the characteristics of the next k request to arrive on the system and simulates the resource management algorithms by using the MatchmakeSched, MMS module to determine if additional resources are to be acquired in the future. It decides to acquire resources if

(a) No available resources satisfy the SLA of one or more of the upcoming k predicted requests

AND

The respective resource acquisitions lead to the prediction of a profit for the broker.

OR

(b) Rejecting some of the upcoming predicted requests leads to a B that is higher than the specified GoS (B_{spec}).

The algorithms for the reactive and proactive components are similar to the reactive and proactive auto-scaling algorithms described in the respective previous sections. A detailed description of a hybrid auto-scaling system is provided in Biswas et al. (2017).

5.4.1 Performance Analysis of the Hybrid Auto-Scaling Algorithm

Simulation-based performance analysis for the algorithm (referred to as System I) was used to compare it with two other alternate systems, System II and System III that were described in Section 5.3.2. Measurements made on a proof-of-concept prototype running on the Amazon EC2 cloud showed close agreements with the simulation results (Biswas et al. 2017).

The description of the workload parameters used in the simulation experiments and the simulation environment is beyond the scope of this chapter. A detailed description of these is available in Biswas et al. (2017). As in the case of proactive and reactive auto-scalers, both BP and UC decrease with an increase in f (see Figures 5.11 and 5.12). The rationale for such a behavior is the same as that described in the respective sections that focused on the performance analysis of the reactive and proactive auto-scaling algorithms. Once again, as in the case of proactive and reactive auto-scaling discussed earlier UC for any given f is the lowest for System I followed by that System II and System III.

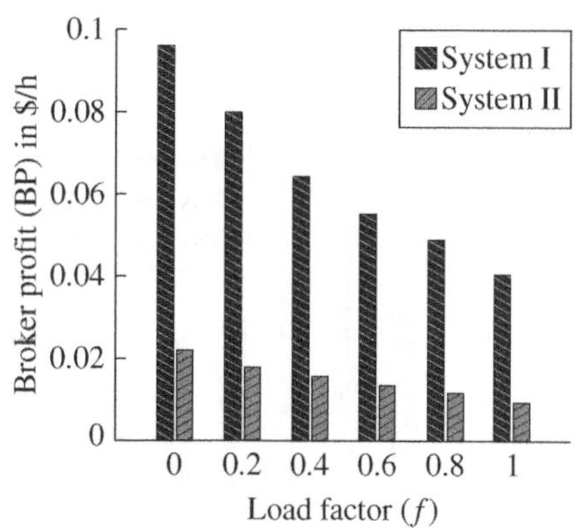

Figure 5.11 Analysis of broker profit achieved by the hybrid auto-scaler. Source: Biswas et al. (2017)/Springer Nature/CC BY 4.0.

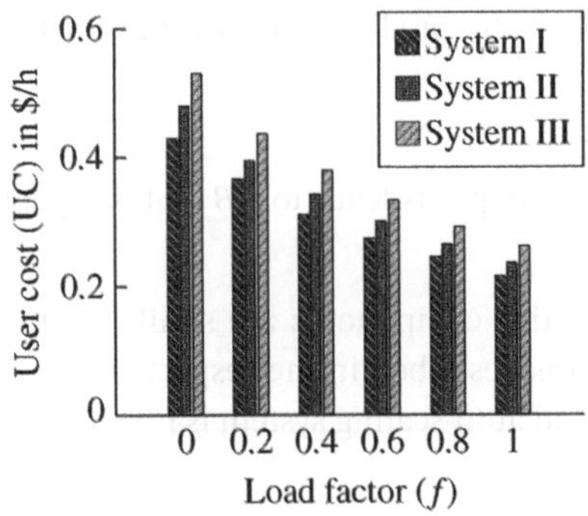

Figure 5.12 Analysis of user costs for the hybrid auto-scaler. Source: Biswas et al. (2017)/Springer Nature/CC BY 4.0.

5.4.2 Comparison with Pure Reactive and Proactive Approaches

The performance of the hybrid auto-scaler (System I) is compared with that of a pure reactive auto-scaler (System I-R) and a pure proactive auto-scaler (System I-P). As shown in Figure 5.13, for any given f, the hybrid auto-scaler leads to a higher broker profit in comparison to the pure reactive and proactive auto-scalers. As shown in Figure 5.14 user cost for the hybrid system is the highest. This is because the hybrid auto-scaler accepts a higher number of requests per unit time in comparison to the two other auto-scalers. Since user cost is directly proportional to the number of requests accepted, a higher number of requests accepted by the hybrid auto-scaler leads to a higher UC.

5.5 Summary

The workload intensity on a typical computer system changes dynamically with time. Provisioning for the peak load intensity leads to poor-resource utilization when the system load is lower, whereas provisioning a lower number of resources

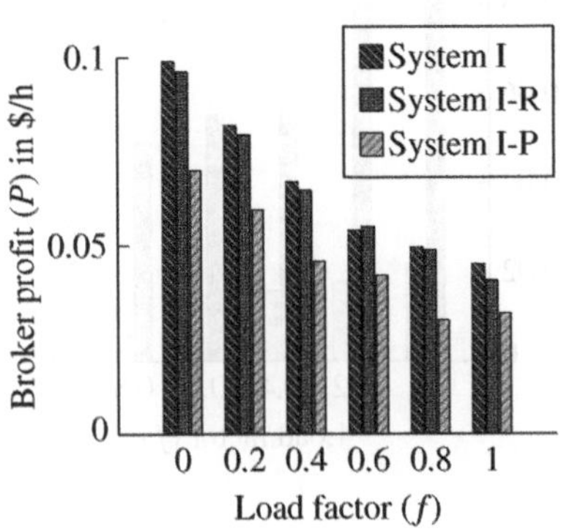

Figure 5.13 Comparison of broker profit. Source: Biswas et al. (2017)/Springer Nature/CC BY 4.0.

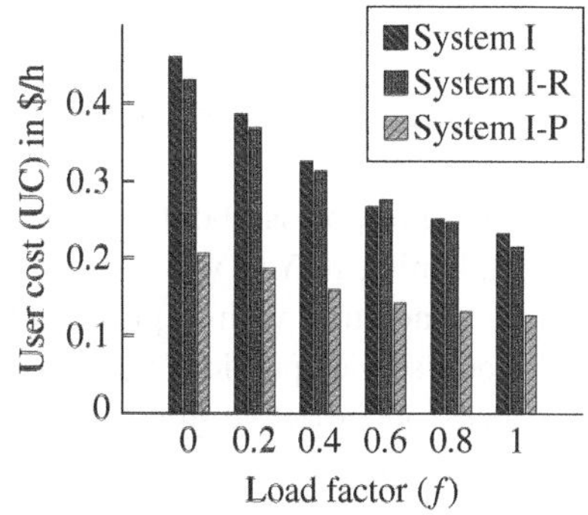

Figure 5.14 Comparison of user cost. Source: Biswas et al. (2017)/Springer Nature/CC BY 4.0.

that is adequate for handling the low load situation can lead to a poor quality of experience for the users when the load is higher. This chapter concerns automatic resource provisioning (auto-scaling) that can change the capacity of resources deployed on the system to match the current workload intensity. A reactive, a proactive, and a hybrid auto-scaling techniques were described. In reactive auto-scaling, the number of resources is changed after a change in load intensity occurs whereas a proactive auto-scaler uses predictive techniques to forecast the future workload and acquires resources in advance if needed to handle the change in workload that is expected to occur in the future. A hybrid auto-scaling technique combines the advantages of both proactive and reactive auto-scaling by deploying both a reactive and proactive auto-scaling system components in its algorithm.

Exercises

1 Describe the following types of auto-scaling:
   a. Horizontal auto-scaling
   b. Vertical auto-scaling
   c. Reactive auto-scaling
   d. Proactive auto-scaling
   e. Hybrid auto-scaling

2 Compare the reactive and proactive auto-scaling approaches described in the chapter.

3 Proactive and reactive approaches have been discussed in the context of horizontal auto-scaling. Is it possible to perform them in the context of vertical auto-scaling? If so, discuss high-level approaches for these techniques. Include the overheads associated with each in your discussion.

4 Describe the role of the padding factor used in the context of the reactive auto-scaling technique described in Section 5.3.1.1. How should one go about in choosing its value?

5 This question is about interpreting and processing experimental data and explaining it. You will have to deal with an auto-scaling system. You will be concerned with a private cloud that gets its resources from a public cloud service provider that was described in the chapter. The problem will focus on proactive auto-scaling. A detailed description of the auto-scaling algorithm/system is provided in Biswas et al. (December 2014).

You are provided with one set of experimental results. In each experiment, a single parameter was varied and the other parameters are held at their default values (see Table 5.1). Each of these parameters is described in Biswas et al. (December 2014). You will need to process the given set of experimental results (see Table 5.2), draw graphs, and discuss the results commenting on the effect of the parameters f (load factor) and ST (mean service time) on the performance of the auto-scaling technique (as indicated by broker profit).

Table 5.1 Default parameter values for the experiments.

Parameter	Default value
Load factor (LF)/f	0.5
Higher arrival rate (HAR) in requests per second/λ_{high}	0.0533
Total number of requests (REQ)	6000
Mean service time (ST) in seconds	3000
Laxity factor (Lax)	1
Number of predictions (PRED)	10
Lower arrival rate (LAR) in requests per second/λ_{low}	0.0267
Public cloud charge (C_PUB/y) in cunits/h	3
User charge (C_PVT/x) in cunits/h	4
Machine learning algorithm (MLA)	LR
Randomize (RANDOM)	NORAND
Training ratio (TRAINING)	0.4

Note: cunit is the currency unit. Further details as to whether it is CDN$, US$, Japanese ¥, or anything else is not important in the context of this problem. The Broker Profit and User/Public cloud charges are also expressed using the same currency unit.

Table 5.2 Each entry in columns 2–4 is the BP for the given combination of f and ST.

ST	f = 0.2	f = 0.5	f = 0.8
2000	0.215895	0.189045	0.181575
3000	0.33975	0.32616	0.30573
4000	0.43305	0.43308	0.38499
5000	0.61149	0.57972	0.58611

Impact of S and f on BP for the System:

(a) Plot a bar graph with BP on the Y-axis and f on the X-axis. There will be four set of bars in the graph one corresponding to a specific value of ST.

(b) Explain the impact of f and S on BP achieved.

6 Consider the profit-based proactive auto-scaling system discussed in this chapter. For simplicity, assume that only one request is predicted by the broker at a time. The broker (intermediary enterprise) pays 2 cents per hour for a resource to the cloud service provider. On the contrary, a client (user) is charged 3 cents per hour for a resource. Note, however, that once the broker acquires a resource from the public CP it needs to pay 2 cents for an hour even if the resource is used for less than an hour. The user, on the contrary, is only charged for the time (in seconds) for which the resource is used.

For the case described below determine whether or not the proactive broker will decide to acquire a resource. Assume that the predicted request's deadline cannot be satisfied on any of the existing resources (that are already acquired by the system).

If the broker decides NOT to acquire a resource compute the minimum increase in the charge for the user such that acquiring a resource becomes profitable for the broker.

Characteristics of the predicted request:

Earliest Start time = 10:00 a.m.; Deadline = 11:30 a.m.; Execution Time = 10 min

7 **Refer** to the system described in question 6 and the following characteristics of the predicted request:Earliest Start time = 10:00 a.m.; Deadline = 1:00 p.m.; Execution Time = T min. Given: $T < 60$ min.

What is the maximum value of T for which the cost incurred by the user for getting this request served by using this broker-based system is lower than the cost that the user will incur if it did not use the broker and got the resource directly from the public CP?

8 Consider a system in which the broker can choose among multiple cloud service providers each with its unique charge. How should the algorithms described in this chapter be modified to handle such an environment?

References

Amazon, "Amazon CloudWatch," n.d.-a, http://aws.amazon.com/cloudwatch/. [accessed July 2019].

Amazon, "Amazon EC2 Pricing," n.d.-b, http://aws.amazon.com/ec2/pricing/. [accessed July 2019].

S. Bhagavathiperumal, and M. Goyal, "Dynamic Provisioning of Cloud Resources Based on Workload Prediction," *Lecture Notes in Networks and Systems (Springer)*, Vol. 75, Issue 1, 2019, pp. 41–49.

A. Biswas, S. Majumdar, B. Nandy, A. El-Haraki, "Automatic Resource Provisioning: A Machine Learning based Proactive Approach", *Proc. IEEE Intl. Conference on Cloud Computing Technology and Science (CloudCom)*, Singapore, December 2014.

A. Biswas, S. Majumdar, B. Nandy, A. El-Haraki, "An Auto-scaling Framework for Controlling Enterprise Resources on Clouds", *Proc. 15th IEEE/ACM International Symp. on Cluster Cloud & Grid Computing (Intl. Workshop on Cloud for Business, Industries and Enterprises (C4BIE))*, Shenzen , May 2015a.

A. Biswas, S. Majumdar, B. Nandy, A. El-Haraki, "Predictive Auto-scaling Techniques for Clouds Subjected to Requests with Service Level Agreements", *Proc. 11th IEEE World Congress on Services (SERVICES)*, June 2015b.

A. Biswas, S. Majumdar, B. Nandy, A. El-Haraki, "A Hybrid Auto-scaling Technique for Clouds Processing Applications with Service Level Agreements", *Journal of Cloud Computing: Advances, Systems, and Applications (Springer)*, Vol. 6, Issue 29, December 2017.

X. Dutreilh, N. Rivierre, A. Moreau, J. Malenfan, I. Truck, "From Data Center Resource Allocation to Control Theory and Back," *Proc. of 3rd IEEE Conference on Cloud Computing (CLOUD), Miami, FL, U.S.A*, July 2010, pp. 410–417.

I. Foster, C. Kesselman, C. Lee, B. Lindell, C. Nahrstedt, A. Roy, "A distributed resource management architecture that supports advance reservations and co-allocation," *Proc. of Seventh International Workshop on Quality of Service (IWQoS)*, London, May 1999, pp. 27–36.

Z. Gong, X. Gu, J. Wilkes, "PRESS: PRedictive Elastic ReSource Scaling for cloud systems," *Proc. of International Conference on Network and Service Management*, Niagara Falls, Canada, October 2010, pp. 9–16.

T. Lorido-Botran, J. Miguel-Alonso, J. A. Lozano, "A Review of Auto-scaling Techniques for Elastic Applications in Cloud Environments," *Journal of Grid Computing*, Vol. 4, Issue 12, 2014, pp. 559–592.

L.R. Moore, K. Bean, T. Ellahi, "Transforming reactive auto-scaling into proactive auto-scaling," *Proc. of the 3rd International Workshop on Cloud Data and Platforms (CloudDP)*, New York, U.S.A, April 2013, pp. 7–12.

RightScale, "RightScale Cloud Management", n.d., https://www.rightscale.com/about. [accessed: July 2019].

J. Yu, R. Buyya, C. K. Tham, "Cost-based scheduling of scientific workflow applications on utility grids," *Proc. of First International Conference on e-Science and Grid Computing*, Melbourne, July 2005, pp. 140–147.

6

Resource Management for Systems Running MapReduce Jobs

6.1 Introduction

This chapter focuses on resource management in systems subjected to a workload comprising data and compute-intensive jobs. As pointed out by multiple sources, large volumes of data are produced every day. For example, according to a Forbes article published in 2018, the volume of data produced daily is of the order of quintillion bytes (Marr 2018). With the growing popularity of the Internet of Things (IoTs) technology, the number of data-producing Internet-enabled mobile devices are also on the rise. Such data, also referred to as BigData, is often stored for analysis. Such bulk or batch data analytics is important in various contexts that include enterprise applications as well as smart cyber–physical systems such as sensor-based bridges and smart buildings (Majumdar 2016). Using single-processor-based conventional systems is often inadequate for processing such data and compute-intensive workloads in a reasonable amount of time and using parallel processing techniques is required. MapReduce is a parallel framework proposed by Google (Dean and Ghemawat 2004) for improving the execution time for such BigData analytics applications. Although MapReduce provides the necessary framework for parallel processing an effective resource management technique is required for using the underlying computing and storage resources on a cloud or cluster processing a workload comprising MapReduce applications. This chapter focuses on resource management on systems executing such MapReduce jobs. An introduction to the MapReduce framework is provided first. It is followed by a discussion of resource management techniques for handling MapReduce applications.

The topics covered in this chapter are briefly described. Section 6.2 introduces the MapReduce framework. Section 6.3 focuses on resource management for MapReduce jobs that are to be completed on a best effort basis. Resource management for systems that are subjected to MapReduce job requests with a

Resource Management on Distributed Systems: Principles and Techniques,
First Edition. Shikharesh Majumdar.

client-specified service level agreement (SLA) is described in Sections 6.4 and 6.5, each of which focuses on a different resource management approach. Section 6.4 presents the description of a budget-based algorithm as well as its performance analysis. A constraint programming (CP)-driven resource management technique is presented in Section 6.5, which also includes a performance analysis of the technique for various combinations of system and workload parameters. Errors are often associated with task execution time estimates that are specified by the users as part of the SLA. Both overestimation and underestimation of the execution times are possible, leading to SLA violations. Techniques for mitigating the impacts of such errors are described in Section 6.6.

6.2 MapReduce

The MapReduce model is based on the splitting of a large data file into smaller chunks and the processing of these chunks in parallel (see Figure 6.1). As shown in the figure, the MapReduce model uses three different phases of execution. In the map phase, multiple map tasks (referred to as Mappers in Figure 6.1) are run concurrently on the system, with each map task processing the data in a given chunk or split. The outputs of the map tasks are then organized in a shuffle phase and passed on to the Reduce tasks (referred to as Reducers in Figure 6.1). The role of the shuffle phase is to group similar data together before passing it on to the Reduce tasks. Note that the specific functions executed by a map task and the logic for performing the grouping in the shuffle phase are application-dependent and are provided by the application designer. The Reduce phase corresponds to the execution of multiple concurrent Reduce tasks that combine and process the data organized in the shuffle phase into the multiple parts of the output data. Once again, the exact operations performed by Reduce tasks are application dependent and vary from one application to another. Irrespective of the exact nature of the applications, a common set of resource management techniques is available in the literature to allocate processing resources to the different tasks and schedule the execution of a set of tasks allocated to a given resource. MapReduce is a parallel processing framework that has been deployed on clouds to facilitate BigData analytics (Dittrich and Quiane-Ruiz 2012) including processing web analytics applications, scientific applications, social networks, and enterprise data. Some data analytics applications can be processed on a best effort basis. That is, the resource manager attempts to minimize the execution latency but does not need to provide any guarantee on the completion time (Collins 2011). For other applications such as live business intelligence, personalized advertising, spam/fraud detection, and real-time analysis of event logs, a given MapReduce job submitted by a user needs to be completed by a deadline provided by the user

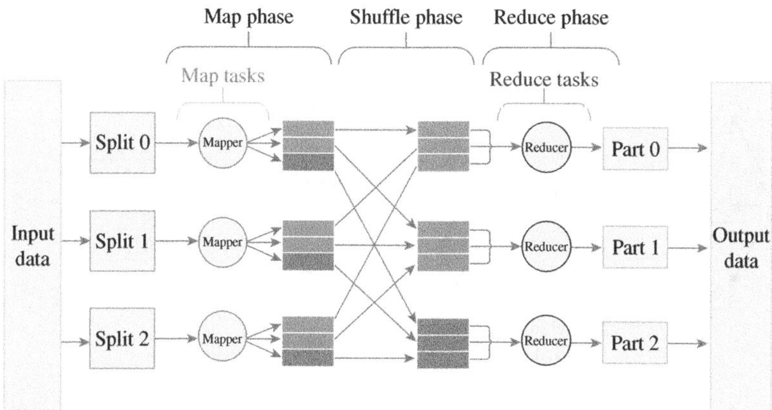

Figure 6.1 The MapReduce model. Source: Lim and Majumdar (2017)/Taylor and Francis.

(Verma et al. 2012). Resource management techniques for systems processing MapReduce jobs on a best effort basis are discussed in Section 6.3, whereas the resource management techniques for systems running deadline-based MapReduce applications are presented in Section 6.4.

The functionality of the MapReduce framework is explained with the help of the well-known word count example (Dean and Ghemawat 2004). The objective of the application is to compute the number of occurrences of each distinct word in an input text file. Consider a text file comprising the following three lines:

Greg Ajit Srinivas
Mary Greg Mary
Srinivas Mary Ajit

The input file is split into three segments, each of which comprises a single line that is handled by a separate map task (see Figure 6.1). The first map task produces a set of intermediate key-value pairs: (Greg, 1); (Ajit 1); (Srinivas, 1). Each key is a distinct word whereas the value field holds the number of times the key appears in the segment (also referred to as split) processed by the first map task. The set of intermediate key-value pairs produced by the second map task is (Mary, 2); (Greg, 1) whereas the set of intermediate key value pairs produced by the third map task is (Srinivas, 1); (Mary, 1); (Ajit, 1). In the shuffling phase, the outputs of the map tasks are combined to produce four sets of key list pairs Greg (1,1); Ajit (1, 1); Srinivas (1,1); Mary (1, 1, 1). Each key corresponds to a distinct word and the associated list indicates the number of times the word appeared in the output of the map tasks. Finally, the lists generated in the shuffling phase are consolidated by the

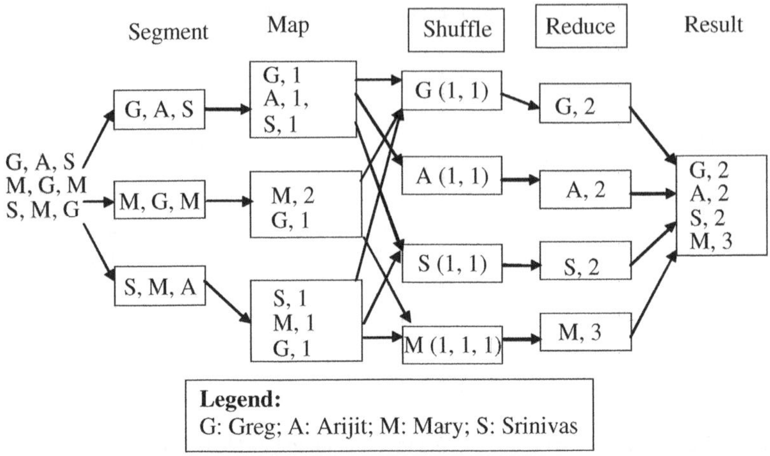

Figure 6.2 Word count example for MapReduce.

Reduce tasks to produce the number of times each word occurs in the input text file: (Greg 2); (Ajit, 2); (Srinivas 2); (Mary, 3). The final result of the MapReduce applications is shown in the last box of Figure 6.2.

6.3 Resource Management Techniques for MapReduce Job Requests to be Satisfied on a Best Effort Basis

This section focuses on resource managers that execute jobs on a best effort basis. That is, the resource manager tries to minimize latency but does not guarantee any time for job completion. Apache Hadoop is a popular implementation of the MapReduce framework (Apache n.d.). A short description of Apache Hadoop and some of its popular schedulers is briefly described next.

A Hadoop cluster has a master–slave architecture. The master nodes perform control operations for the cluster, whereas the slave nodes execute the Map and Reduce tasks. The NameNode is responsible for the file system name space where as a JobTracker node is responsible for distributing the jobs among the worker nodes. In addition to the nodes executing the tasks, the slave nodes include a TaskTracker that manages job execution: starting/stopping of jobs, monitoring job status, and informing JobTracker of a job completion. DataNodes are also slave nodes that are used by the distributed file system.

A number of schedulers are available on this platform. Three such schedulers are discussed next.

- **FIFO Scheduler**

This is the default scheduler for the system. Jobs upon arrival join a queue. Jobs for execution are selected from this queue. Jobs are scheduled in the same order in which they arrived. That is, if Job A has arrived before Job B, it is executed prior to Job B on the system.

- **Fair Scheduler**

Fair Scheduler is based on the fairness principle. Its goal is to assign resources to jobs in such a way that all jobs receive, on average, an equal share of resources over time. By default, the Fair Scheduler focuses primarily on the memory resource. However, it can be configured to schedule with both memory and CPU (Hadoop n.d.-a).

- **Capacity Scheduler**

The capacity scheduler is for multitenant large clusters. Each tenant is associated with a person or organization and is associated with a minimum resource capacity for running Map and Reduce tasks. The capacity scheduler provides the additional benefit that a tenant can access any excess capacity not being used by others. This provides the capability of resource sharing to the organizations in a cost-effective manner. Job priority is supported by the capacity scheduler. A higher-priority job is provided access to system resources earlier in comparison to lower-priority jobs. More details on the capacity scheduler are available in (Hadoop n.d.-b).

Scheduling of MapReduce jobs has received attention from researchers. Two classes of such algorithms, *data locality aware algorithms,* and *optimization technique-based algorithms* are discussed in the literature. Algorithms in the first class have used locality aware job placement such that data transfer between a storage device and a computing machine is reduced. For example, the technique described in (He et al. 2011) tries to schedule a task on a node that already contains the input data for the tasks. Another example is the Locality-Aware Reduce Task Scheduler (LARTS) described in (Hammoud and Sakr 2011). LARTS considers both the size and the location of input data for a task and tries to place tasks on nodes in such a way that unnecessary network traffic is minimized. The second class of algorithms includes a number of resource managers that have used techniques from the system optimization literature for performance enhancement. Linear Programming based techniques, for example, have been described in (Chang et al. 2011; Gao et al. 2012) for reducing job completion times in a cluster running MapReduce jobs. The well-known CP-based optimization techniques have been utilized in scheduling jobs with deadlines that is discussed in a following section.

6.4 Resource Management Techniques for MapReduce Job Requests with Service Level Agreements

As discussed earlier, many data analytics applications deploying the MapReduce framework are latency sensitive. A request submitted by users of such MapReduce jobs is associated with an SLA. The SLA comprises (i) the earliest start time for the job, (ii) estimates of task execution times, and (iii) a deadline by which the job must be completed. Note that each component of the SLA is provided by the user submitting the job. The resource manager needs to allocate the resources for the tasks in the job and schedule the start times for the tasks in such a way that the user-specified deadline for job completion is met.

Resource management for job requests to be satisfied on a best effort basis is already a well-known NP-hard problem. Associating a deadline for completion makes the job of the resource management algorithm even harder. Two algorithms for resource management for MapReduce jobs with deadlines are described in the two following Sections 6.4.1 and 6.5.

6.4.1 The Budget-Based MapReduce Resource Management Technique

The overhead of running optimization techniques such as CP-based tools (described in a later section) may often restrict the scalability of the system. A budget-based resource management technique that does not use CP may be useful at higher values of arrival rates that cannot be sustained by the CP-based scheduler. Such an algorithm and the corresponding resource manager MapReduce Budget-Based Resource Manager (MRBB-RM) are discussed in (Lim et al. 2014a). The algorithm is presented in this section. Although MRBB-RM is more scalable, at low-to-medium arrival rates, the proportion of missed deadlines achieved by this algorithm is often inferior to that achieved by a CP-based algorithm called MapReduce CP-based Resource Manager (MRCP-RM) which is discussed later.

6.4.1.1 High-Level Description

MRBB-RM comprises three components:

- Algorithm 1 (the Job and Task Mapping Algorithm)
- Algorithm 2 (the Deadline Budgeting Algorithm)
- Algorithm 3 (the Job Remapping Algorithm)

The relationship among these three algorithms is captured in Figure 6.3.

The main Algorithm 1 is described in Figure 6.4. A job upon arrival invokes Algorithm 1, which in turn invokes Algorithm 2. Algorithm 1 is responsible for mapping the tasks in the arriving job onto the available resources such that the job's SLA is met. To achieve its goal, it generates a set of sub-deadlines for

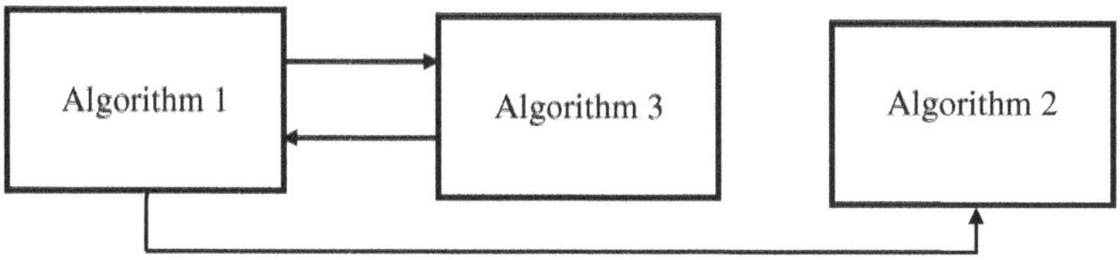

Figure 6.3 Components of MRBB-RM.

the tasks in the job from the deadline for the job submitted as part of its SLA by the user. Algorithm 2 is responsible for performing this operation: decomposing the overall job deadline into sub-deadlines for its constituent tasks. If Algorithm 1 is unable to develop a schedule in which all the tasks of the job can meet their deadlines, leading to a deadline violation for the job Algorithm 3 is invoked. The responsibility of Algorithm 3 is to remap the tasks of jobs that can potentially affect the completion time of the job missing its deadline and to perform remapping of the tasks in the jobs so that the job can meet its deadlines. Algorithm 3, while performing the remapping uses the assistance of Algorithm 1. A short description of the main algorithm (Algorithm 1) is presented next. The description of the operations performed by Algorithm 2 and Algorithm 3 are explained in the context of the description provided for Algorithm 1. Further details of these algorithms are available from (Lim et al. 2014a).

Algorithm 1 is invoked by the arrival of a job j on the system. Upon invocation, Algorithm 1 calls Algorithm 2 budgetDeadline (line 1), which divides the slack time available for the job and distributes it among the Map and Reduce tasks. Note that the slack time of a job of is the amount of time the job has between its deadline and the end of the job's execution if it starts executing at its earliest start time. The tasks in job j are then sorted in nondecreasing order of sub-deadlines (line 2). Thus, resources are allocated first to tasks with more imminent deadlines. For each task in job, j the algorithm allocates the task on the resource on which the task can be executed at the earliest possible time (lines 3–4). If the task cannot be allocated on a resource to meet the deadline of j, then Algorithm 3, remapJob, is called (lines 5–8). Algorithm 3 tries to reallocate tasks on the system that have not started execution yet such that the deadline of j can be met without violating the deadline requirements of all the jobs the resources for which have already been allocated. If the reallocation attempt is successful or if all the tasks in job j have been successfully mapped, the job is added to a mapped jobs list (lines 10–12). Otherwise, job j is mapped without any consideration for its deadline (lines 14–16) although it may lead to a deadline miss for j.

A simulation-based performance analysis for the budget-based technique is presented in (Lim et al. 2014a) and is summarized in the following section.

Algorithm 1: Job and Task Mapping Algorithm, *mapJob()*

Input: Job to map *j*; Boolean *isRootJob*;
Output: True if job is successfully mapped, false otherwise.

```
call budgetDeadline(j) if j's deadline has not been budgeted
taskToMap ← Get j's tasks and sort by non-decreasing order of
sub-deadlines (earliest sub-deadline first).
for each task t in tasksToMap
    Find resource rs in set of resources R that executes t at the
    earliest possible time, and map t on rs.
    if t cannot be mapped before j's deadline then
        call remapJob(j, isRootJob, 1) returning
        (Boolean) isJobRemapped
        break
    end if
end for
if isJobRemapped = true or all j's tasks mapped then
    Add j to mappedJobs list
    return true
else
    if isRootJob = true then
        Map j ignoring deadline and return true
    end if
    return false
end if
```

Figure 6.4 Algorithm 1. Source: Lim et al. (2014a)/with permission of IEEE.

6.4.2 Performance Analysis

The job arrivals in the simulated system follow a Poisson process. The number of Map and Reduce tasks in a job, task execution times are generated based on a uniform distribution. Further discussion of the workload and system parameters used in the simulation is beyond the scope of this chapter. A detailed description of these parameters used in the performance analysis and the methods used in their generation are available in Lim et al. (2014a). The paper includes a detailed performance analysis of MRBB-RM. Only a representative set of experiments are discussed here to provide insights into the execution behavior of the algorithm and its performance.

The two performance metrics discussed in this section are presented next.

Proportion of Late Jobs (P): is the ratio of the number of jobs that missed their deadlines and the total number of jobs considered in the simulation experiment. The lower the value of P smaller is the number of SLA violations and better is the system performance.

Average Job Turnaround Time (T): The turnaround time for a job is the difference between its completion time and its earliest start time provided as part of the SLA by the user. A shorter turnaround time typically leads to a higher degree of user

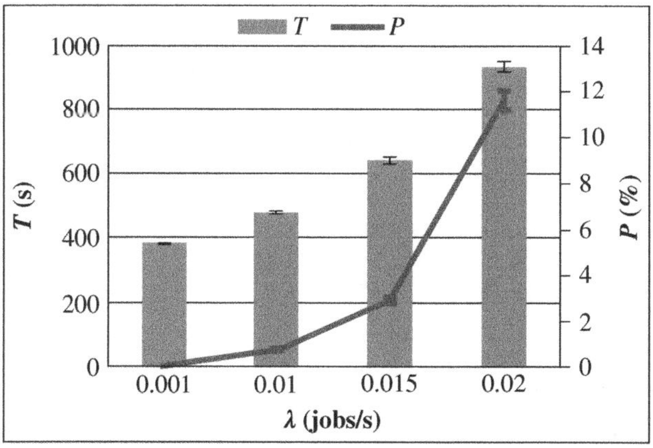

Figure 6.5 Effect of arrival rate of jobs on performance. Source: Lim et al. (2014a)/with permission of IEEE.

satisfaction. The average job turnaround time is the average of the turnaround times of all the jobs considered in a given experiment.

The Effect of Job Arrival Rate: As the job arrival rate (λ) increases more jobs arrive on the system per unit time. This gives rise to an increase in system contention leading to a higher queuing delay. As a result, T increases with λ (see Figure 6.5). An increase in system contention giving rise to delays in job execution leads to an increase in the proportion of jobs that miss their deadlines. This is captured in the increase with P with an increase in the arrival rate of jobs as displayed in Figure 6.5.

The Effect of Task Execution Times: The impact of e_{max} the upper bound for the uniform distribution used in generating the map task execution times is presented in Figure 6.6. A higher value of e_{max} indicates a higher execution time for map tasks. The increase in execution times for map tasks has a similar impact on system performance as the increase in arrival rate. As e_{max} increases system contention increases leading to an increase in both T and P (see Figure 6.6).

The second resource management algorithm for MapReduce jobs with SLAs is described in the following section.

6.5 The Constraint Programming-Based MapReduce Resource Management Technique

A CP-based resource management algorithm, *MapReduce CP-Based Resource Manager (MRCP-RM),* that can aptly handle an open stream of MapReduce jobs with SLAs is presented in this section.

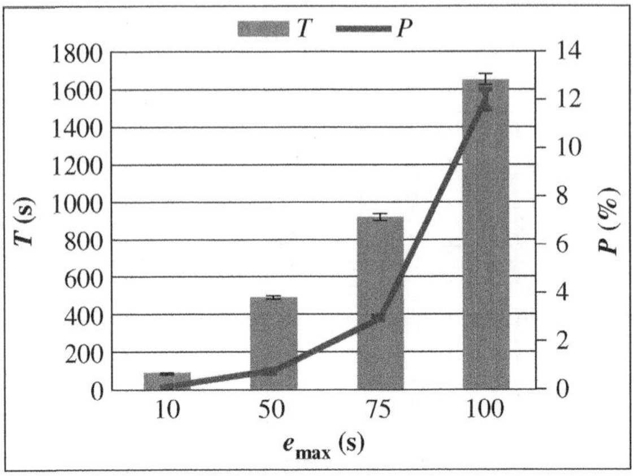

Figure 6.6 Effect of task execution time on performance. Source: Lim et al. (2014a)/with permission of IEEE.

Users submit jobs that are enqueued on the system. Each job is associated with a user-specified SLA that (as described in Section 6.4) comprises the earliest start time, estimates of execution times for Map and Reduce tasks, as well as a deadline by which the job is to be completed. A high-level description of the different components of MRCP-RM is presented in Figure 6.7. When the resource manager, MRCP-RM is free (i.e. it is not busy performing resource management for other jobs), it performs resource allocation and scheduling. In addition to the newly

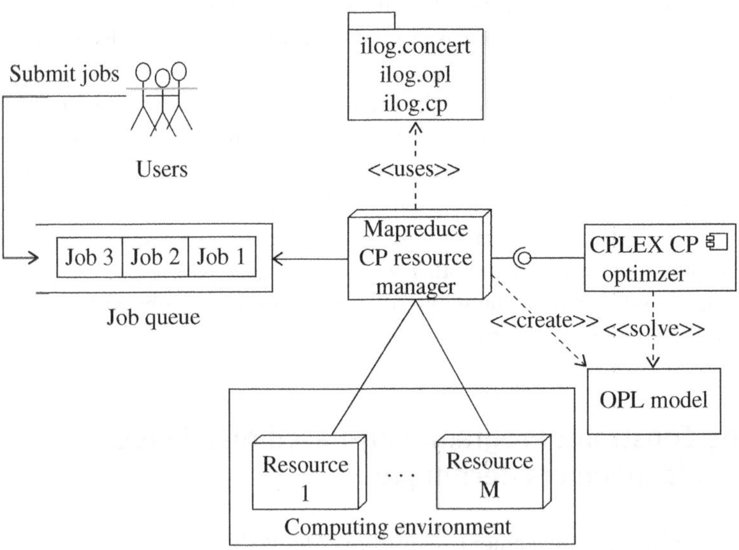

Figure 6.7 Description of the components of MRCP-RM. Source: Lim et al. (2014b)/with permission of IEEE.

arrived jobs all the other jobs that have not yet started executing are considered during this resource management process. Reconsidering the previously arrived jobs improves the flexibility in scheduling that, in turn, minimizes the number of late jobs.

The matchmaking and scheduling are formulated as a CP model that comprises constraints for each task in the MapReduce jobs that are being considered by the resource manager. MRCP-RM uses IBM CPLEX (IBM Corporation 2021) to generate an Optimization Programming Language (OPL) model by using IBM's OPL for describing the constraints. The OPL model is then solved by the CPLEX optimizer engine. Once the model is solved the mapping of tasks to resources and their schedule of execution become available. That is, MRCP-RM determines which resource (1 … *M*) is to be used for running a given task and the time at which the task will start running.

The decision variables of the CP model are described:

- x_{tr}: A binary variable used in *matchmaking*. It is set to 1 if a task *t* is assigned to a resource *r*; otherwise, x_{tr} is set to 0. Note that x_{tr} variables exist for *all* combinations of task *t* in set *AT* (described later) and resource *r* in set of all resources *R*.
- a_t: An integer variable used for *scheduling*. It specifies the assigned start time of a task *t*. Note that there is an a_t variable for each task *t* in set *AT*.
- N_j: A binary variable that is set to 1 if job *j* misses its deadline. Note that each job *j* in the set of all jobs *J* has an N_j variable that is initialized to 0.

The set of constraints is described in Figure 6.8.

Figure 6.8 The CP model. Source: Lim et al. (2017a)/with permission of IEEE.

$$Minimize \sum_{job_j \in J} N_j \quad \text{such that}$$

$$\sum_{res_r \in R} x_{tr} = 1 \quad \forall\, t \in AT \tag{6.1}$$

$$(a_t mp \geq s_j \quad \forall t^{mp} \in T_j^{map}) \quad \forall job_j \in J \tag{6.2}$$

$$\left(a_t rd \geq \max_{t^{mp} \in T_j^{mp}} (a_t mp + e_t mp) \quad \forall\, t^{rd} \in T_j^{rd}\right) \forall job_j \in J \tag{6.3}$$

$$\left(\max_{t^{rd} \in T_j^{rd}} (a_t rd + e_t rd) > d_j \Rightarrow N_j = 1\right) \forall job_j \in J \tag{6.4}$$

$$\begin{pmatrix} cumulative((a_t|x_{tr} = 1), (e_t|x_{tr} = 1), \\ (q_t|x_{tr} = 1), c_t^{mp}) \quad \forall\, t \in T^{mp} \end{pmatrix} \forall res_t \in R \tag{6.5}$$

$$\text{where } T^{mp} = \mathrm{U}_{jobj \in J} (T_j^{mp})$$

$$\begin{pmatrix} cumulative((a_t|x_{tr} = 1), (e_t|x_{tr} = 1), \\ (q_t|x_{tr} = 1), c_t^{rd}) \quad \forall t \in T^{rd} \end{pmatrix} \forall res_T \in R \tag{6.6}$$

$$\text{where } T^{rd} = \mathrm{U}_{jobj \in J} (T_j^{rd})$$

$$(x_{tr} \in \{0, 1\} \quad \forall\, t \in AT) \quad \forall res_r \in R \tag{6.7}$$

$$N_j \in \{0, 1\} \quad \forall job_j \in J \tag{6.8}$$

$$a_t \in \mathbb{Z} \qquad \forall\, t \in AT \tag{6.9}$$

The objective function of the CP model is the minimization of the number of late jobs. The first constraint ensures that each task is mapped to a single resource only: x_{tr} is 1 for only one value of r and 0 for the others. Constraint (2) ensures that the start times of all map tasks in every job j in J are greater than or equal to the respective earliest start times S_j for the job included in the SLA. Constraint (3) ensures that the Reduce tasks of job j start only after all the map tasks of j have been completed. Constraint (4) updates the value of N_j when job j completes. The completion time of the job is the completion time of the last finishing Reduce task in the job. It is given by the sum of its start time a_t^{rd} and execution time e_t^{rd}. If this completion time is later that its deadline d_j then the job is marked as late, and N_j is set to 1. Constraints (5) and (6) ensure that the numbers of simultaneously running Map and Reduce tasks on any resource at any given point in time do not exceed the respective Map and Reduce slot capacity of the resource, respectively. These constraints are formulated using the CP global constraint function, *cumulative*. Each resource res_r gives rise to two such constraints one for the map tasks (Constraint 5) and one for the Reduce tasks (Constraint 6). The cumulative constraint computes the number of executing tasks of a particular type (Map or Reduce) and enforces that this number at any point in time is less than or equal to the resource capacity limit for the particular task type at each resource. For a particular constraint, only the tasks that are assigned to that resource (i.e., $xtr = 1$) are included as input during the evaluation of the cumulative function used in that constraint. Lastly, the remaining Constraints (7)–(9) define the domain of the decision variables: xtr, N_j and a_t specifying the valid values that the respective variables can take. The MRCP-RM algorithm that uses this set of constraints is described next (Lim et al. 2017a).

6.5.1 The MRCP-RM Algorithm

The algorithm (see Figure 6.9) referred to as "Algorithm 1" in (Lim et al. 2017a)) starts off by updating the earliest start times of tasks (lines 1–3). For each resource res_r in the set of resources in R all the tasks allocated to the resource are checked (lines 5 and 6). If the task should start later, it is skipped (lines 7 and 8). Otherwise, it implies that the task has stared executing. If the task's end time is greater than the current time a new constraint specifying the task's scheduled start and end times and the resource on which it is running is added to the OPL model (line 11) and the status of the task stored in the PrevScheduled field is set to be true (line 12). On the contrary, if the task's end time is less than or equal to the current time then the task t is deemed to be complete and is removed from the parent job's tasks list (line 14). Once all the tasks in a job have completed execution the job is removed from the set of jobs, J (line 16). The next step is to create a new OPL model and attach the input data comprising J and R (line 19). The model is thus generated

Algorithm 1. MRCP-RM

```
Input: A set of jobs J to map on a set of Resources R.
1:  for each job_j in J (j = 1..n) do
2:      if j's earliest start time is less than the current time then
3:          Set j's earliest start time to the current time.
4:  end for
5:  for each res_r in R (r = 1..m) do
6:      for each scheduled task t in res_r do
7:          if t's start time is greater than the current time then
8:              break
9:          else
10:             if t's end time is greater than the current time then
11:                 Add a new constraint to the OPL Model that
                    specifies the start time, end time, and assigned
                    resource of t.
12:                 Set t's isPrevScheduled field to true.
13:             else
14:                 Record that t is complete and remove it
                    from its job's tasks list.
15:                 if the job's tasks list is empty then
16:                     Remove the job from J.
17:     end for
18: end for
19: Create a new OPL Model and attach the data source
    containing J and R.
20: Generate and solve the OPL Model.
21: When a solution is found, extract and save the values of
    the decision variables (assigned start time and assigned
    resource) for each task.
```

Figure 6.9 The MRCP algorithm. Source: Lim et al. (2017a)/with permission of IEEE.

and solved (line 20). When a solution is obtained the values of the scheduled start time and task to resource mapping for all the tasks on the system are saved in the last step (line 21).

6.5.2 Performance Analysis for MRCP-RM

Performance analysis of MRCP-RM as captured in Figures 6.10 and 6.11 is presented.

An implementation of MRCP-RM on a Hadoop cluster that is referred to as Hadoop CP-Based Resource Manager (HCP-RM) is described in (Lim et al. 2017a). The performance of the CP-based technique (MRCP-RM) as captured by HCP-RM is compared to that of the well-known earliest deadline first (EDF) technique (see Figure 6.10). EDF associates a higher priority with jobs that have

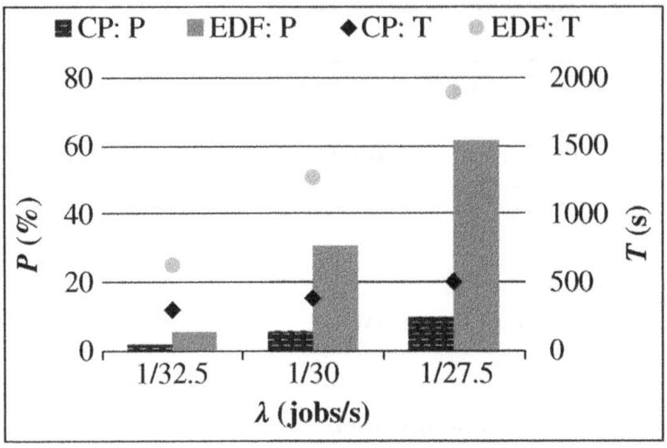

Figure 6.10 HCP-RM (CP) versus EDF-scheduler (EDF): effect of job arrival rate on P and T. Source: Lim et al. (2017a)/with permission of IEEE.

shorter deadlines. Further details of EDF and the description of the experiments performed on an Amazon EC2 system are described in detail in (Lim et al. 2017a). The experimental system is subjected to a synthetic workload that allows the varying of various systems and workload parameters. Two performance metrics, the proportion of jobs that miss their deadlines (*P*) and the average turnaround time of jobs (*T*) shown for CP and EDF are captured in Figure 6.10. The figure presents the results of the experiments in which the arrival rate of jobs, λ, is varied. As λ increases, the contention for resources increases in the system, and both P and T are observed to increase. For any given λ, The CP-based algorithm outperforms EDF by a large margin for both the performance metrics. Figure 6.11 presents the effect of varying λ on a system subjected to the execution of a real MapReduce application "Wordcount" that counts the number of times a word appears in a document (Lim et al. 2017a). A similar relative performance between MRCP-RM (as captured by HCP-RM) and EDF is observed. Further details of the experiments including the size of the document used are available from (Lim et al. 2017a). The figures demonstrate the superiority of CP over EDF.

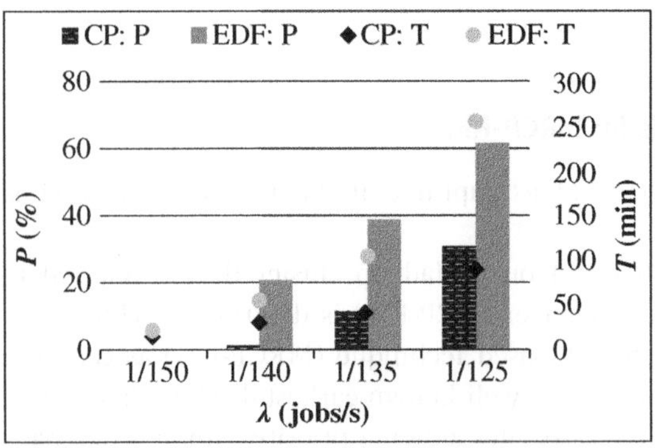

Figure 6.11 HCP-RM (CP) versus EDF-scheduler (EDF): effect of arrival rate on *P* and *T* when using the WordCount workload. Source: Lim et al. (2017a)/with permission of IEEE.

6.6 Errors Associated with User Estimates of Task Execution times

Estimation of task execution times is hard and there is often an error/inaccuracy associated with the execution times of tasks estimated by users that are sent as part of the SLA. The previous history of execution of the same job can be used for predicting task execution times. Code profilers are also used for estimating job run times. Irrespective of the method used existing literature shows that the estimates of job run times are error prone (Bailey Lee et al. 2004).

Both overestimation and underestimation of task execution times can be detrimental to system performance. With an overestimated task execution time the task completes before the time that has been determined by the resource manager. This tends to introduce idle time on a resource that was booked for the task thus lowering its utilization. With an underestimated task execution time, the task remains incomplete after the expiry of the time interval for which the resource was booked for its execution leading to a potential abortion of the job. Such an abortion tends to reduce the useful utilization of the resource. Building robustness into resource management techniques for handling such inaccuracies is thus important for clusters and clouds.

The incorporation of error-handling techniques into MRCP-RM is discussed next. Two types of error-handling mechanisms described in the literature are discussed. A *prescheduling error-handling technique* (PSEH) (Lim et al. 2017b) that is directed at generating new estimates of task execution times based on the estimates submitted by the user is presented in the next subsection. The prescheduling technique is applied before a job starts running on the system. A *runtime error-handling technique* (RTEH) (Lim et al. 2017c) that handles tasks with overestimated/underestimated execution times after the job has started execution is presented in the following section.

6.6.1 The Prescheduling Error-Handling Technique

This technique is based on using the history of past job submissions for predicting the inaccuracy associated with the execution times of tasks in the job that just arrived on the system. After a job completes the actual execution times of tasks are known and the error factor, f_t, associated with the estimated task execution time submitted by the user is computed as:[1]

$$f_t = \left(e_t^{\text{est}} - e_t^{\text{run}}\right) / e_t^{\text{run}} \quad (6.10)$$

1 Equation numbering continues with equation 6.10 that follows equation number (9) in Figure 6.8.

where e_t^{run} is the actual run time of task t that was measured when the job was running on the system and e_t^{est} is the execution time estimate for task t submitted as part of the SLA for the job by the user. Note that the system keeps statistics on the average task error factor (tef) for all map tasks and all Reduce tasks. The average map task error factor is the average of the f_t values for all map tasks that have been completed on the system, whereas the average reduce task error factor is the average of the f_t values for all Reduce tasks that have been completed on the system.

After a job arrives, the system uses tef to generate an adjusted task execution time:

$$e_t^{\text{adj}} = e_t^{\text{est}} / (1 + \text{tef}) \tag{6.11}$$

Note that depending on whether task t is a map or a Reduce task the respective tef computed for Map or Reduce tasks is used.

A detailed description of the algorithm for PSEH is presented in (Lim et al. 2017b).

6.6.1.1 Performance Analysis

A prototype implementation of the main MRCP algorithm on Hadoop called the Hadoop CP-based resource manager, HCP-RM, is compared with HCP-RM-EH that incorporates the PSEH error-handling technique into HCP-RM. Experiments were performed on a Hadoop cluster comprising 1 master and 10 slave nodes running on an Amazon EC2 cloud to compare the performance of the two techniques. A synthetic workload that generated the number of Map and Reduce tasks and the various task execution times using uniform distributions was used. A detailed description of the experimental setup is available in (Lim et al. 2017b). Three performance metrics, P, the proportion of jobs that missed their deadlines, T, the average job turnaround time and O, the system overhead due to the CPU time consumed in running the resource management algorithm are used in the performance analysis. Three different error models from the literature were used in (Lim et al. 2017b) to generate errors associated with user-estimated task execution times. The experimental results demonstrate the superiority of HCP-RM-EH that outperformed HCP-RM in terms of all three performance metrics. The variation in P with error factor f used in the constant error model (England et al. 2005) is captured in Figure 6.12. With the constant error model the estimated user estimated task execution time, e_t^{est}, is related to the actual task execution time, e_t, as follows:

$$e_t^{\text{est}} = e_t(1 + f) \tag{6.12}$$

For each task execution time, $e_{t, \text{produced}}$ by the uniform distribution used in the synthetic workload generator, the estimated task execution time, e_t^{est}, is generated by using Eq. (6.12). As shown in Figure 6.12, HCP-RM-EH demonstrates a

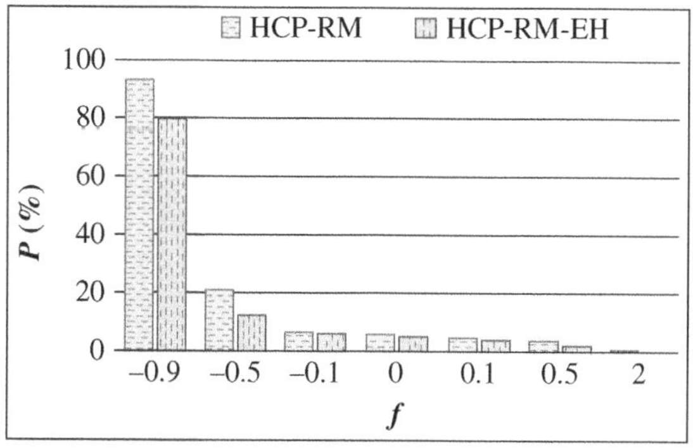

Figure 6.12 HCP-RM versus HCP-RM-EH: effect of f on P when using the constant error model. Source: Lim et al. (2017b)/with permission of IEEE.

superior performance: for any value of f, the proportion of jobs that miss their deadlines (P) achieved by HCP-RM-EH is smaller than that achieved with HCP-RM. The missing bar for HCP-RM-EH at $f = 0.2$ indicates a $P = 0$ achieved by the error-handling technique.

6.6.2 The Runtime Error-Handling Technique

As in the case of PSEH, RTEH is devised to handle both jobs with overestimated task execution times and those with underestimated task execution times. The algorithm is invoked in two situations: (i) when a task has been completed before its scheduled end time (that occurs when the task execution time was overestimated) and (ii) when the task is not completed by its scheduled end time (that occurs when the task execution time was underestimated). In case (i) the completed task is removed from the system and the tasks that are scheduled to execute on the resource slot used by the completed task at later points in time are started earlier if the earliest start time requirements for the tasks are not violated. The idle resource times for the future created by moving the execution of these tasks forward in the resource schedule are typically filled by new jobs arriving on the system. This tends to reduce resource idle times and improve system performance. In case (ii), the incomplete task is given additional time quanta to complete its operation as long as this does not lead to deadline misses for other jobs. In the method described in (Lim et al. 2017c), two options are considered. With Underestimation Error-Handling Mechanism 1 (UE1), tasks of all jobs with underestimated task execution times are provided with additional time. With Underestimation Error-Handling Mechanism 2 (UE2) only those jobs that have not missed their deadlines are provided with additional time quanta for completion.

6.6.2.1 Performance Analysis

A simulation-based performance analysis for RTEH is reported in (Lim et al. 2017c). The job arrivals follow a Poisson process. The number of Map and Reduce

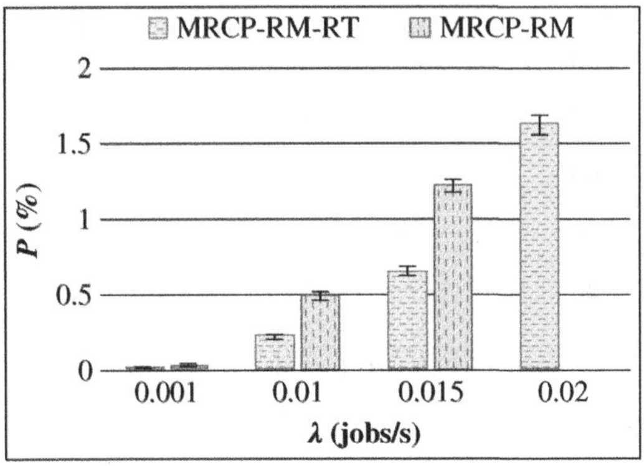

Figure 6.13 Performance comparison for various arrival rates: *P*. Source: Lim et al. (2017c).

tasks in a job, task execution times are generated based on a uniform distribution. A complete list of workload and system parameters used in the performance analysis is available in Lim et al. (2017c). The constant error model with $f = 0.5$ was used. A positive value of f implies that the execution times of all tasks are overestimated. The proportion of late jobs, P, the average job turnaround time T, and the resource management overhead associated with running the algorithm O are computed (see Figures 6.13 and 6.14). As the job arrival rate, λ, increases the contention for resources increases leading to an increase in P for both MRCP-RM, and MRCP-RM-RT (see Figure 6.13). As shown in Figures 6.13 and 6.14, for any given arrival rate MRCP-RM-RT demonstrates a superior performance in comparison to MRCP-RM for all the three performance metrics, P, T, and O. This shows the effectiveness of RTEH to make more efficient use of resources when the task execution times are overestimated. Note that the absence of results for MRCP-RM when λ is 0.2 jobs per sec is because MRCP-RM was not usable to handle this high arrival rate and the system ran out of memory before the

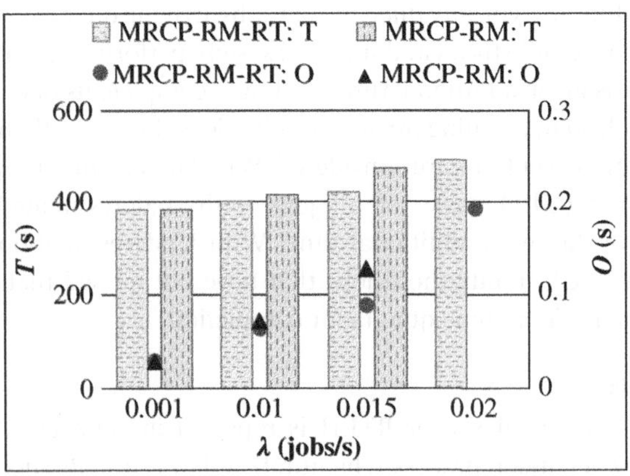

Figure 6.14 Performance comparison for various arrival rates: T and O. Source: Lim et al. (2017c).

completion of the experiment. This is an indication of the higher scalability of MRCP-RM-RT in comparison to MRCP-RM.

Simulation results for underestimated task execution times show that UE1 and UE2 give rise to comparable performance (Lim et al. 2017c).

6.7 Summary

This chapter concerns resource management on platforms running MapReduce jobs that are used in data analytics as well as for other applications. The focus of the chapter was on the allocation and scheduling of jobs associated with SLAs that included a deadline for completion. Two techniques for resource management, a budget-based technique, and a CP-based technique, were discussed. The SLA associated with a job includes user estimates of task execution times that are often error prone. Two techniques for handling such errors and increasing the robustness of the MRCP-RM algorithm were discussed.

Exercises

1 Compare the different features of the MRCP-RM algorithm and the MRBB-RM algorithm. Discuss the relative merits and demerits of the two algorithms, including the overheads associated with each.

2 Consider the MRCP-RM algorithm. Explain the role of each constraint described in Figure 6.8.

3 The MRCP-RM algorithm described in the chapter is for systems that are subjected to an open stream of AR requests. [That is, the requests arrive one by one on the system]. How will you adapt the algorithm for a system processing a batch workload comprising multiple ARs that arrive at the same time on the system? Assume that each job has its own deadline.

4 Repeat question 3 for a situation in which all the ARs in the system have the same deadline.

5 Repeat question 3 with the MRBB-RM algorithm.

6 Repeat question 4 with the MRBB-RM algorithm.

7 Consider a system in which jobs have a priority associated with them. Discuss the approach for adapting (a) MRCP-RM and (b) MRBB-RM for such a system.

References

Apache, Hadoop Wiki, n.d., http://wiki.apache.org/hadoop.

C. Bailey Lee, Y. Scwartzman, J. Hardy, A. Snavely, "Are User Runtime Estimates Inherently Inaccurate?", in *Job Scheduling Strategies for Parallel Processing* (eds: D. G. Feitelson, L. Rudolph, and U. Schwiegelshohn), Berlin, New York, USA, 2004, pp. 253–263

H. Chang, M. Kodialam, R.R. Kompella, T.V. Lakshman, M. Lee, S. Mukherjee, "Scheduling in MapReduce-Like Systems for Fast Completion Time", *Proc. of IEEE INFOCOM 2011*, Shanghai, China, 10–15 April 2011, pp. 3074–3082.

M. Collins, "Hadoop and MapReduce: Big Data Analytics," *Gartner*, January 14, 2011, https://www.gartner.com/en/documents/1521016.

J. Dean, S. Ghemawat, "MapReduce: Simplified Data Processing on Large Clusters", *Proc. of International Symposium on Operating System Design and Implementation (OSDI)*, San Francisco, CA, USA, 6–8 December 2004, pp. 137–150.

J. Dittrich, J.-A. Quiane-Ruiz, "Efficient Big Data Processing in HadoopMapReduce", *Proc. of VLDB 2012/PVLDB*, Vol. 5, no. 12, Istanbul, Turkey, August 2012, pp. 2014–2015.

D. England, J. Weissman, J. Sadagopan, "A New Metric for Robustness with Application to Job Scheduling", *Proc. of Internatioanl Symposium on High Performance Distributed Computing*, 24–27 July 2005, pp.135–143.

X. Gao, Q. Chen, Y. Chen, Q. Sun, Y. Liu, M. Li, "A Dispatching-Rule-Based Task Scheduling Policy for MapReduce with Multi-type Jobs in Heterogeneous Environments", *Proc. ChinaGrid Annual Conference*, Beijing, China, 20–23 September 2012, pp. 17–24.

Hadoop, "Fair Scheduler", n.d.-a, https://hadoop.apache.org/docs/r2.7.4/hadoop-yarn/hadoop-yarn-site/FairScheduler.html. [accessed 2023].

Hadoop, "Hadoop: Capacity Scheduler", n.d.-b, https://hadoop.apache.org/docs/r1.2.1/capacity_scheduler.html. [accessed 2023].

M. Hammoud, M.F. Sakr, "Locality-Aware Reduce Task Scheduling for MapReduce", *Proc. of International Conference on Cloud Computing Technology and Science (CloudCom)*, Athens, Greece, 29 November–1 December 2011, pp. 570–576.

C. He, Y. Lu, D. Swanson, "Matchmaking: A New MapReduce Scheduling Technique", *Proc. of International Conference on Cloud Computing Technology and Science (CloudCom)*, Athens, Greece, 29 November–1 December 2011, pp. 40–47.

IBM Corporation, "IBM ILOG CPLEX Optimization Studio V12.5", 2021.

N. Lim, S. Majumdar, "Resource Management for MapReduce Jobs Performing Big Data Analytics", in *Big Data Management and Processing* (eds: K.-C. Li, H. Jiang, and A. Zomaya), Chapman & Hall/CRC Press, May 2017.

N. Lim, S. Majumdar, P. Ashwood-Smith, "Resource Management Techniques for Handling Requests with Service Level Agreements", *Proc. of 2014 International*

Symposium on Performance Evaluation of Computer and Telecommunication Systems (SPECTS), Monterey, USA, July 2014a.

N. Lim, S. Majumdar, P. Ashwood-Smith, "Constraint Programming-Based Resource Management Technique for Processing MapReduce Jobs with SLAs on Clouds", *Proc. of International Conf. on Parallel Processing (ICPP)*, Minneapolis, USA, October 2014b.

N. Lim, S. Majumdar, P. Ashwood-Smith, "MRCP-RM: a Technique for Resource Allocation and Scheduling of MapReduce Jobs with Deadlines", *IEEE Transactions on Parallel and Distributed Systems*, Vol. 28, Issue 5, 4 May 2017a.

N. Lim, S. Majumdar, P. Ashwood-Smith, "Techniques for Handling Error in User-estimated Execution Times during Resource Management on Systems Processing MapReduce Jobs", *Proc. of IEEE/ACM International Conference on Cluster, Cloud and Grid Computing (CCGrid)*, Madrid, Spain, May 2017b.

N. Lim, S. Majumdar, P. Ashwood-Smith, "A Run Time Technique for Handling Error in User Estimated Execution Times on Systems Processing MapReduce Jobs with Deadlines", *Proc. of The IEEE 5th International Conference on Future Internet of Things and Cloud (FiCloud)*, Prague, Czech Republic, August 2017c.

S. Majumdar, "Cloud-Based Smart Facilities Management", in *Internet of Things Principles and Paradigms* (ed: D. Buyya, Elsevier, 2016, pp. 319–339.

B. Marr "How Much Data Do We Create Every Day? The Mind-Blowing Stats Everyone Should Read", May 2018, https://www.forbes.com/sites/bernardmarr/2018/05/21/how-much-data-do-we-create-every-day-the-mind-blowing-stats-everyone-should-read/?sh=4603a94a60ba [accessed: March 2024]

A. Verma, L. Cherkasova, V.S. Kumar, R.H. Campbell, "Deadline-Based Workload Management for MapReduce Environments: Pieces of the Performance Puzzle," *Proc. Network Operations and Management Symposium*, Maui, Hawaii, USA, 16–20 April 2012, pp. 900–905.

7

Energy Aware Resource Management

7.1 Introduction

A significant amount of power is consumed by data centers that are used in providing Information Technology (IT) support to enterprises and other institutions. The cost of power consumption is an important component of the operational cost of a data center. A high power consumption may also lead to higher carbon dioxide emissions. Reducing the server power consumption, which is a contributor to the overall power consumption for a data center can also lead to a reduction in cost due to the power consumed by the cooling infrastructure used in data centers. The total aggregate power cost of a data center has often been estimated as being a significant component of the monthly operational cost of the center. Reducing the power consumption of data centers is important and is driving the invention of lower-cost components as well as energy aware resource management techniques. This chapter continues the discussion on resource management techniques for systems subjected to MapReduce jobs presented in Chapter 6 focusing on energy aware resource management. The techniques discussed are devised for not only meeting the client-specified service level agreements (SLAs) but they also attempt to reduce the power consumption of servers that run various IT applications.

Dynamic Frequency and Voltage Scaling (DVFS) is a technique invented for controlling the power consumed by the server. This technique is based on the impact of clock frequency on the power consumed by a CPU. Reducing the clock frequency reduces the CPU operating frequency that reduces the speed of instruction execution and decreases the power consumed by the device. The clock frequency can be conveniently controlled by adjusting the voltage supplied to the CPU. A standard Application Programming Interface (API) called the Advanced Configuration and Power Interface (ACPI) is available for performing this operation (Hewlett-Packard Corporation 2011). In fact, by using DVFS with

Resource Management on Distributed Systems: Principles and Techniques,
First Edition. Shikharesh Majumdar.

ACPI it is possible for the operating system to control the power consumed by every core in an ACPI-compliant multicore CPU.

Another technique that is often used for reducing energy consumption is based on consolidating several executing tasks on a subset of the total number of servers and shutting off the idle servers created by this consolidation operation. Once new tasks arrive, and more capacity is needed one or more of these idle servers are woken up and the existing tasks can be redistributed among these active servers on the system if needed. Each of these techniques is discussed in the following sections.

The DVFS technique is discussed first in Section 7.1.1. The server-consolidation-based energy management technique is described later in the chapter. Resource management for both open systems subjected to a stream of MapReduce job arrivals as well as systems running a given batch of jobs is discussed. A technique called Energy Aware MapReduce Resource Manager (EAMR-RM) for managing resources on open systems subjected to a continuous arrival of MapReduce jobs with SLAs is presented in Sections 7.2 and 7.3. A detailed performance analysis of EAMR-RM capturing the impact of various system and workload parameters on performance is reported in the various subsections of Section 7.3. A configurable resource manager for a batch system running a given batch of MapReduce jobs is described in Section 7.4 while its performance analysis in presented in Section 7.5. As mentioned earlier another approach to energy savings is to reduce the number of active servers when possible and is discussed in Section 7.6. A summary of the key issues described in this chapter is presented in Section 7.7.

7.1.1 Dynamic Voltage Frequency Scaling

This technique is based on the relationship between CPU power, voltage, and operating frequency for the CPU. As discussed in the previous paragraph using the API specified in the ACPI standards the power consumed by an ACPI-compliant CPU can be controlled dynamically even during the execution of programs. Such CPUs typically support a given number of voltage and frequency levels. The frequency/voltage of the CPU is thus changed to alter its power consumption. The maximum and minimum frequency levels that can be used with the CPU will be denoted by f^{high} and f^{low} respectively. Although a given CPU can directly support a discrete set of frequency voltage levels, any desired average operating frequency can be achieved by controlling the proportion of time for which a specific frequency from the given range is used.

The total power, P, dissipated by a CPU is the sum of two components: static power dissipation P_{s} and dynamic power dissipation P_{d}:

$$P = P_{\text{s}} + P_{\text{d}}$$

A number of researchers (see Wang et al. 2010 for example) have shown that static power is proportional to dynamic power:

$$P_s = kP_{d,}$$

where k is the proportionality constant. The total power can thus be expressed in terms of dynamic power only:

$$P = (1 + k) P_d$$

It has been shown that the proportionality constant k depends only on the system architecture and is not dependent on the program under execution (Piguet et al. 2004). P_d, in turn, can be expressed as a function of the percentage of active gates (A), the total capacitance load (cap), the supply voltage, V, and the operating frequency, f (Halimi et al. 2013) for the CPU:

$$P_d = A * \text{cap} * V^2 * f$$

It has been demonstrated that the cap depends on the machine architecture. Existing work (Halimi et al. 2013) shows that A can be reasonably approximated as being a constant for the duration of the execution of a given program. Thus, the power P consumed for a given program is computed as:

$$P = \delta V^2 f,$$

where δ is a constant. Therefore, if the program executes for a duration of $\Delta\tau$ for a given frequency and voltage combination, f and v, the total energy consumed by the CPU during the execution of the given program can be computed as:

$$E = \sum_{\Delta\tau} P * \Delta\tau.$$

A number of power and sleep modes that are referred to as C-states are supported by ACPI-compliant CPUs (Hewlett-Packard Corporation 2011). An idle CPU core is allowed to enter one of the sleep modes C1–C6 and switched back to the operating state C0. C1 is referred to as the halt state and can be used as the sleep mode since the transition time between C1 and C0 is negligibly small in comparison to typical execution times for programs. A halted CPU core typically uses a proportion (referred to as h in this discussion) of the power consumed by it when operating at a frequency of f^{high} that corresponds to the operating state C0. Using deeper sleep states (C2–C6) can reduce the idle power but can increase the time required for transitioning from the idle to the active state.

The DVFS-based technique can be used on a machine that runs in isolation or in the context of virtual machines that run on a cloud environment. When a specific VM's frequency of operation needs to be changed the change can be directly applied to the core if each VM is running on a separate core. If VMs are created by time slicing a CPU the change in frequency can be applied during a context switch that corresponds to the switching of operation to the respective VM that starts running on the CPU.

7.2 DVFS-Based Resource Management Techniques

The DVFS-based technique can be integrated with various matchmaking and scheduling algorithms. This section presents a DVFS-based resource management algorithm for an environment running MapReduce jobs. MapReduce is a well-known framework used in the context of BigData applications that often require the use of multiple computing machines for long periods of time, leading to a significant energy consumption in the data center running such applications.

An introduction to the MapReduce framework and a resource management technique for meeting client-specified deadlines were discussed in Chapter 6. The discussion presented in the subsequent subsections of this chapter focuses on resource management on systems subjected to a MapReduce workload with the primary objective of minimizing energy consumption while avoiding the missing of the client-specified job deadlines. A number of existing resource management techniques have focused primarily on reducing job completion times for either lowering the average job turnaround time on the system or minimizing the number of deadline misses. Completing jobs at the earliest possible time is not necessarily the best solution for a data center because minimization of job turnaround times implies running the CPUs at their full speed that will increase the power consumed by the servers while running the applications. Reducing the frequency of operation will reduce power consumption but will tend to run the jobs for a longer period of time that may lead to an increase in the total energy consumption for the data center. A careful selection of operating frequency is thus required to minimize the energy consumption for data centers, while complying with the job deadlines that are important components of the client SLAs. A resource manager called EAMR-RM (Gregory and Majumdar 2016a) is discussed next. A high-level diagram for the resource management system is presented in Figure 7.1.

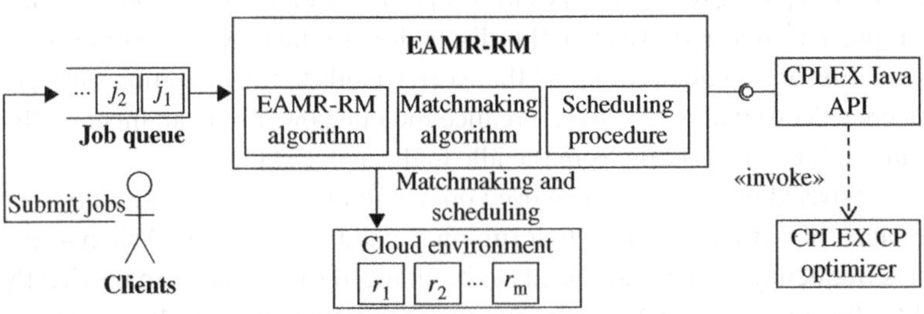

Figure 7.1 The energy aware resource management system. Source: Gregory and Majumdar (2016a)/with permission of IEEE.

As in the case of MRCP-RM described in Chapter 6, the submitted MapReduce jobs join a job queue. When the resource manager is not busy it runs the matchmaking and scheduling algorithm for all the jobs currently in the queue by running the EAMR-RM algorithm. The main objective of the algorithm is to reduce server energy consumption while meeting the deadlines of the accepted jobs. It is important to comply with the client SLA after accepting the respective client request. In the rare event in which the deadline for a job cannot be met the system rejects the job submitted by the client and conveys this information to the client. As in Chapter 6, a constraint programming-based approach using the CPLEX CP optimizer is used by the algorithm. In order to respond to changes in workload and system state EAMR-RM can reallocate and schedule tasks to different resources. Once a task has started execution, it is not preempted. Moreover, the algorithm does not reallocate a task or change the schedule of the task that has already started execution or will start executing within the expected solve time of the CP optimizer. In order to control system overhead the execution time for the optimizer for a given run is fixed.

The generation of the constraint model and the algorithm for EAMR-RM are described in the following sections.

7.3 The EAMR-RM Algorithm

Before discussing the EAMR-RM algorithm the constraint formulation used by the algorithm is presented (see Figure 7.2).

The decision variables used in the constraint list presented in Figure 7.2 are described first. This is followed by the description of the constraint program that uses these variables.

- Decision variable a_t: is an integer variable that specifies the start time of task t
- Decision variable c_t: is an integer variable that specifies the execution duration of task t. Note that execution durations will vary based on CPU frequency

The objective function for the constraint program CP is the minimization of the total energy consumed for executing all the tasks in T (see Figure 7.2). The energy required for the execution of a given task t is given by:

$$E_t = P(e_t/c_t * f^{\text{high}}) * c_t,$$

where $P(x)$ is the function that determines the CPU power used when the operating frequency is x. The value of x results from the observation that e_t and c_t are, respectively, the task execution time submitted by the user as part of the SLA, and that resulted after frequency scaling. The total task energy is the sum of E_t for all t in T.

$$minimize \sum_{t \in T} P(\frac{e_t}{c_t} f^{high}) * c_t + h * P(f^{high}) * (\sum_{r \in R} (\max_{j \in J}(d_j) * (c_r^{map} + c_r^{red}) - \sum_{t \in T} x_{tr} * c_t)) \; subject \; to$$

$$\sum_{r \in R} x_{tr} = 1 \quad \forall t \in T \tag{1}$$

$$(a_t \geq s_j \forall t \in T_j^{map}) \forall j \in J \tag{2}$$

$$\left(a_{t'} \geq \max_{t \in T_j^{map}} (a_t + c_t) \forall t' \in T_j^{red}\right) \forall j \in J \tag{3}$$

$$\left(\max_{t \in T_j^{red}} (a_t + c_t) \leq d_j\right) \forall j \in J \tag{4}$$

$$(cumulative((a_t | x_{tr} = 1), (c_t | x_{tr} = 1), 1, c_r^{map}) \forall t \in T_t^{map})) \forall r \in R \tag{5}$$

$$(cumulative((a_t | x_{tr} = 1), (c_t | x_{tr} = 1), 1, c_r^{red}) \forall t \in T_t^{red})) \forall r \in R \tag{6}$$

$$(x_{tr} \in \{0,1\} \forall t \in T) \forall r \in R \tag{7}$$

$$\left(c_t \in \mathbb{Z} \mid e_t \leq c_t \leq e_t \frac{f^{high}}{f_{low}}\right) \forall t \in T \tag{8}$$

$$a_t \in \mathbb{Z} \; \forall t \in T \tag{9}$$

Figure 7.2 List of constraints. Source: Gregory and Majumdar (2016a)/with permission of IEEE.

Constraint (1) states that each task is to be allocated to a single resource. Constraint (2) ensures that the start time of each task is greater than or equal to its earliest start time. Constraint (3) specifies that none of the reduce tasks can start execution before the completion of all map tasks. Constraint (4) is used to enforce that the job deadlines are not missed by ensuring that the finishing time for the latest task in the job j does not exceed the deadline for job j. Constraints (5) and (6) are used to ensure that the aggregate resource capacity of the resource pool for each job type map and reduce, is not exceeded. Constraints (7), (8), and (9) describe the valid domains for the variable: x_{tr}, c_t, and a_t used in the constraint program.

The EAMR-RM algorithm that uses CP is presented next (see Figure 7.3).

The algorithm is run periodically. At the beginning of the current invocation (lines 1–17) the algorithm identifies all the tasks that have already started execution or are expected to start execution before the end of execution of the current iteration of the algorithm. Tasks that do not satisfy any of these criteria are not considered in the current iteration (lines 3–4), and tasks that will be considered in this iteration are identified and marked as *Fixed* (line 6). If the task has not completed executing, its start time and end time are specified by adding additional constraints for the task (line 8). Otherwise, the task is simply removed from the

Input: A set of MapReduce jobs *J* to map and schedule on a set of cloud resources *R*

```
1:   for each r in R
2:     for each task in current schedule mapped to r
       (sorted by increasing start time)
3:       if start time is greater than current time plus maxTime
4:         break
5:       else
6:         set isFixed to true
7:         if end time is greater than current time
8:           add constraints to CP formulation to fix resource
             allocation, start time, and end time of task
9:         else
10:          remove task from the job's task list
11:          if the job task list is empty
12:            remove job from J
13:          end if
14:        end if
15:      end if
16:    end for
17:  end for
18:  for each job j in J and queue
19:    if s_j is less than current time plus maxTime
20:      set s_j to current time plus maxTime
21:    end if
22:  end for
23:  create new CP model based on CP formulation for jobs in J and
     queue, resources in R, and fixed tasks
24:  solve CP formulation using IBM CPLEX CP Optimizer
25:  if a solution is found
26:    remove jobs from queue and add jobs and corresponding tasks
       to current list
27:    extract decision variables from solution and update current
       resource allocation and schedule for all tasks
28:  else
29:    reject and discard arriving job from queue
30:  end if
```

Figure 7.3 EAMR-RM algorithm. Source: Gregory and Majumdar (2016a)/with permission of IEEE.

task list for the job (lines 9–10). If the job does not have any incomplete tasks, it is deemed to have completed and is removed from J (lines 11–12). The start times of all the remaining jobs in J and those that are present in the queue are set to the maximum of their current start time and the sum of the current start time and maxTime: an upper bound on the run time of the CP optimizer used to limit the resource management overhead (lines 18–22). The new CP model for all the jobs in J and the given set of resources R is created in line 23. The Scheduling procedure is then invoked in line 24 to run the CP optimizer and the schedule information in terms of the start time, and end time for every job is extracted in line 27. The matchmaking algorithm is invoked to allocate the tasks to resources.

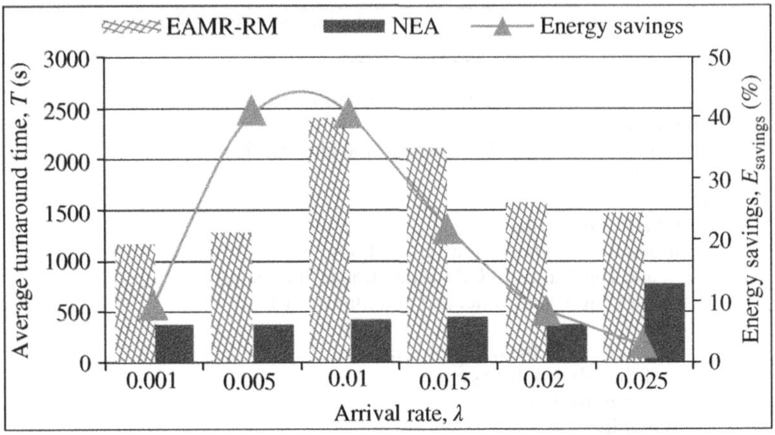

Figure 7.4 The impact of arrival rate on performance. Source: Gregory and Majumdar (2016a)/with permission of IEEE.

If no feasible solution is found by the CP optimizer the arriving job is rejected (line 29).

7.3.1 Performance Analysis of EAMR-RM

A simulation-based performance analysis of the algorithm is presented in Gregory and Majumdar (2016a). The details of the workload parameters used as well as the way they were generated for the simulation experiments are not important in the current context that focuses on the analysis of two important performance metrics. The interested reader can find these details in the paper referenced. Figures 7.4 and 7.5 compare the performance of EAMR-RM with that of a Non-Energy Aware Algorithm (referred to as NEA) which is the same as MRCP-RM discussed in chapter. NEA does not consider the energy consumption due to job execution and focuses only on the minimization of deadline misses by jobs.

The two performance metrics used are described first. *Average turnaround time (T)* is the average time between the arrival and completion of all accepted MapReduce jobs, whereas *Energy Savings* ($E_{savings}$) is the percentage reduction (over NEA) in total energy required to execute an open stream of arriving MapReduce jobs using EAMR-RM.

7.3.1.1 The Impact of Arrival Rate of Jobs

T achieved with NEA is observed to increase with the arrival rate (see Figure 7.4). As more and more jobs arrive per unit time on the system resource contention increases resulting in higher average queuing delays that, in turn, lead to an increase in the average turnaround time. *T* achieved with EAMR-RM displays a nonmonotonic relationship with λ. For lower values of λ, *T* is observed to

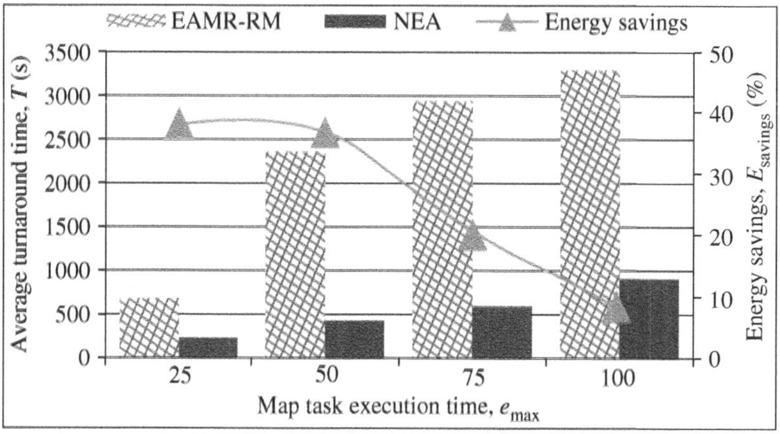

Figure 7.5 The impact of map task execution times on performance. Source: Gregory and Majumdar (2016a)/with permission of IEEE.

increase with the arrival rate. At lower arrival rates the system is able to lower the clock frequency resulting in a higher energy savings, $E_{savings}$. But for higher values of λ resource contention and queuing delays increase as a result of which the algorithm needs to increase the CPU operating frequency such that the user-specified deadlines can be met. This tends to lower the execution time for the jobs and decrease T. Because of the increase in the clock frequency less energy is saved and $E_{savings}$ decreases with an increase in the job arrival rate.

7.3.1.2 The Effect of Map Task Execution Times

The impact of Map task execution times captured by e_{max} the upper bound on the uniform distribution used to model the map task execution times is presented in Figure 7.5. An increase in the arrival rate brings about a change in T for both the algorithms, and the average job turnaround times increase with an increase in task execution times. EAMR-RM needs to increase the CPU operating frequency for higher map task execution times thus reducing the energy savings at higher values of e_{max}.

7.3.1.3 The Impact of Number of Resources

Parallelism of execution increases as the number of resources, m, on which the jobs are executed increases. The relationship of the Average Turnaround T and the Energy Savings, $E_{savings}$ with m are presented in Figure 7.6. Increasing the number of resources speedup execution, and T decreases. The energy savings displayed in the figure are the result of two different phenomena, one caused by changing the CPU clock frequency and the other caused by the energy consumed by idle servers. For a lower number of resources, most of the resources need to be operated at close to full speed as a result of which the energy savings resulting from DVFS used for altering the CPU frequencies of servers is small. $E_{savings}$ increases as m increases.

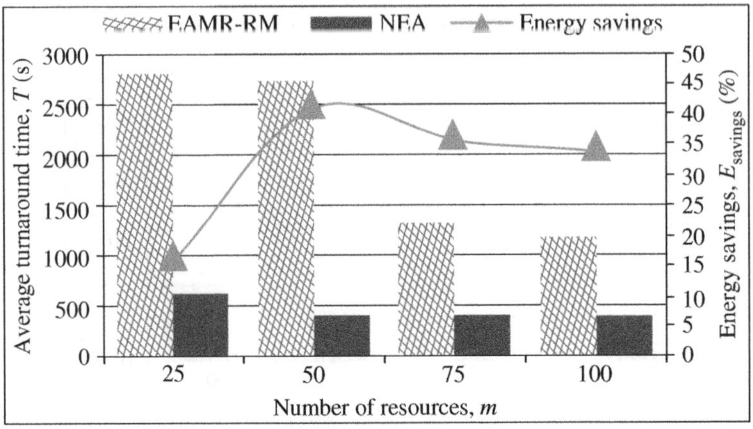

Figure 7.6 Relationship with the number of resources. Source: Gregory and Majumdar (2016a)/with permission of IEEE.

But after a certain value of m is reached, $E_{savings}$ is observed to decrease with an increase in m. This is because at such higher values of m a number of servers tend to remain idle for significant periods of time, and the increase in energy consumed by idle servers seems to offset the energy savings resulting from DVFS.

7.4 Configurable Resource Manager for Processing a Batch of MapReduce Jobs

The EAMR-RM technique for processing an open stream of jobs described earlier focused on minimizing only a single metric: the energy consumed by the system. This often leads to rejecting jobs the SLA requirements for which cannot be met. A system that can reject jobs in such an unconstrained fashion and/or lead to large job turnaround times may not be usable in all environments. This section describes a resource manager that optimizes multiple performance metrics and minimizes energy consumption for executing a given batch of MapReduce jobs while ensuring that client-specified upper bounds on (i) the proportion of jobs that are rejected and on (ii) batch completion times are never violated.

Such a Configurable Energy Aware Resource Manager (CRM) is presented in the following subsection. CRM accepts a batch of MapReduce jobs as well as upper bounds on the missed deadline ratio (defined later) and batch completion time. It then performs resource allocation and scheduling in such a way such that energy consumption is minimized, while the specified missed deadline ratio and batch completion times are not exceeded.

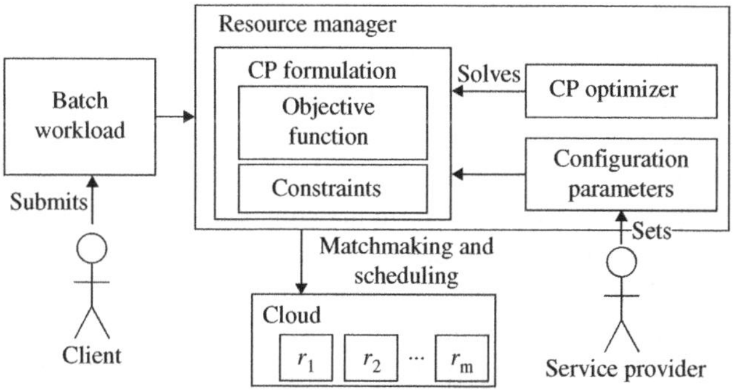

Figure 7.7 The CRM System. Source: Gregory and Majumdar (2016b)/with permission of IEEE.

7.4.1 Constraint Program for the Configurable Resource Manager

The CRM system is explained with the help of Figure 7.7. The user submits a batch of deadline constrained MapReduce jobs along with the SLA parameters that include an upper bound on the missed deadline ratio *db* and an upper bound on batch completion time *cb*. The first provides the maximum proportion of deadline misses for the jobs that are allowed in the batch, whereas the second provides an upper bound on the batch completion time. The execution times for the tasks in each job and the job deadlines are also specified by the user.

A high-level description for the configurable resource manager is provided in Figure 7.7. CRM constructs a constraint program for performing matchmaking and scheduling. The configuration parameters, upper bounds on the missed deadline ratio, and the batch completion time, are used by CRM during the construction of the constraint program. The IBM CPLEX CP optimizer is then used to solve the constraint program and generate the allocation of tasks to the resources and the tasks' start times. The MapReduce jobs are then executed on the cloud using the matchmaking and scheduling information generated by CRM (see Figure 7.7). A short discussion of the system used is provided next. A more detailed description of the system can be found in Gregory and Majumdar (2016b).

The variable T is used here to store all the tasks: $T = \{T_1, T_2, \ldots, T_n\}$. The start time s_j for all jobs is set to the start time for the batch of jobs. R is the set of all the m resources:

$$R = \{r_1, r_2, \ldots, r_m\}.$$

Each resource r, is characterized by the following characteristics: a map task capacity c_r^{map}, a reduce task capacity c_r^{red}. In accordance with the Hadoop model

c_r^{map} and c_r^{red} denote the number of fixed size slots on resource r for executing the map tasks and the reduce tasks, respectively.

Each of these slots is modeled as a CPU core that is associated with a set of permitted frequency/voltage configurations and a halt percentage, h, used in the computation of idle power.

A constraint-based approach is used to perform matchmaking and scheduling. A constraint program for the batch of MapReduce jobs and the resource set R is solved by a CP optimizer. Figure 7.8 presents the list of constraints used by the CP optimizer to generate the allocation and scheduling of tasks. Before explaining the constraints, a brief description of the key variables used in the constraint program is presented.

- x_{tr}: is the binary variable used for all combinations of t in T and r in R. *It* is set to one if task t is assigned to resource r and to zero otherwise.
- a_t: specifies the scheduled start time for task t
- c_t: specifies the execution duration for task t. Note that this execution time is dependent on CPU frequency.
- N_j: is a binary variable used for each job j: *it* is set to zero if j meets its deadline and to one otherwise.

The list of constraints (con1–con12) listed in Figure 7.8 is briefly explained.

Constraint 1: ensures that every task is allocated only a single resource

Constraint 2: is used to signify that each task can only begin execution at or after its earliest start time

Constraint 3: specifies that a Reduce task can only start execution after all the map tasks in the respective job have completed

Constraint 4: indicates that N_j should be set to 1 when the completion time for job is greater than d_j thus violating the deadline specified for j.

Constraints 5 and 6: ensure that the map and reduce task capacities for each resource is not violated.

Constraint 8a: enforces that the proportion of jobs that miss their deadline remains within the bound *db* specified by the user.

Constraint 9a: ensures that the batch completion time remains within the bound *cb* specified by the user.

Constraints 7, 10, 11, and 12: specify the domains of the decision variables

(The letter a used with obj, Constraints 8 and con 9 distinguish them from similar constraints used in an alternate formulation described in Gregory and Majumdar (2016b) and are not important in the current context).

The objective function concerns the minimization of total energy that is the sum of task energy E_{Task}, and idle energy, E_{Idle} each of which is discussed next. Energy required to execute a given task is given by:

$$E_t = P(f^{\text{high}} * e_t/c_t) * c_t,$$

minimize $E_{\text{Task}} + E_{\text{Idle}}$ subject to	**(obja)**
$\sum_{r \in R} x_{tr} = 1 \forall t \in T$	(con1)
$\left(a_t \geq s_j \forall t \in T_j^{\text{map}}\right) \forall j \in J$	(con2)
$\left(a_{t'} \geq \max_{t \in T_j^{\text{map}}} (a_t + c_t) \forall t' \in T_j^{\text{red}}\right) \forall j \in J$	(con3)
$\left(\max_{t \in T_j^{\text{red}}} (a_t + c_t) > d_j \implies N_j = 1\right) \forall j \in J$	(con4)
$(\text{cumulative}((a_t \vert x_{tr} = 1), (c_t \vert x_{tr} = 1), 1, c_r^{\text{map}}) \forall t \in T_r^{\text{map}})) \forall r \in R$	(con5)
$(\text{cumulative}((a_t \vert x_{tr} = 1), (c_t \vert x_{tr} = 1), 1, c_r^{\text{red}}) \forall t \in T_r^{\text{red}})) \forall r \in R$	(con6)
$(x_{tr} \in \{0, 1\} \forall t \in T) \forall r \in R$	(con7)
$\sum_{j \in J} N_j \leq \vert J \vert * db$	(con8a)
$\max_{t \in T} (a_t + c_t) \leq cb$	(con9a)
$N_j \in \{0, 1\} \forall j \in J$	(con10)
$\left(c_t \in \mathbb{Z} \mid e_t \leq c_t \leq e_t \frac{f^{\text{high}}}{f_{\text{low}}}\right) \forall t \in T$	(con11)
$a_t \in \mathbb{Z} \ \forall t \in T$	(con12)

Figure 7.8 CP formulation. Source: Gregory and Majumdar (2016b)/with permission of IEEE.

where *P(x)* is the function that determines CPU power for a CPU operating frequency of x and c_t is the task execution time. The expression for E_{Task} is the sum of energy consumed for all the tasks in T

$$E_{\text{Task}} = \sum_{t \in T} P\left(\frac{e_t}{c_t} f^{\text{high}}\right) * c_t$$

E_{Idle} is a function of P_{Idle} that is computed as:

$$P_{\text{Idle}} = h * P\left(f^{\text{high}}\right)$$

The expression for E_{Idle} is:

$$E_{\text{Idle}} = P_{\text{Idle}} * \left(\sum_{r \in R} (\max_{t \in T} (a_t + c_t) * (c_r^{\text{map}} + c_r^{\text{red}}) - \sum_{t \in T} x_{tr} * c_t)\right)$$

7.5 Performance Analysis of CRM

The performance analysis of CRM based on a simulation model used with a synthetic workload is summarized. Further discussion of the simulation parameters is

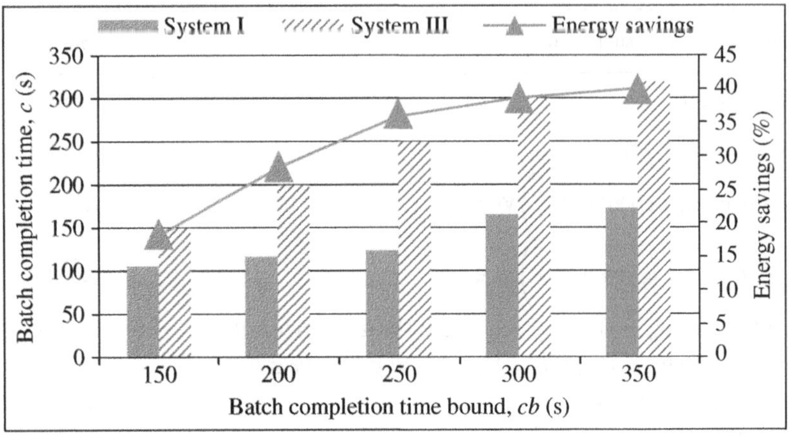

Figure 7.9 Impact of the batch completion time bound. Source: Gregory and Majumdar (2016b)/with permission of IEEE.

beyond the scope of this chapter. A detailed description of the system and workload models is presented in Gregory and Majumdar (2016b).

The performance of CRM (referred to as System II in the figure) is compared to a nonenergy aware constraint programming-based resource manager referred to as System I (Gregory and Majumdar 2016b) that does not perform any energy optimization but simply focuses on satisfying the deadline requirements of jobs.

7.5.1 Effect of Batch Completion Time Bound

The graph presented in Figure 7.9 shows the Batch Completion Time and Energy Savings for different values of the Batch Completion Time Bound. Energy Savings is computed as:

$$E_{\text{savings}} = \left(1 - \frac{E_{(\text{for System II})}}{E_{(\text{for System I})}}\right) * 100\%.$$

For smaller values of *cb* there is less scope for CRM to lower the CPU operating frequency that increases job execution times, and may therefore, lead to a violation of the given bound on batch completion time. Thus, a lower value of E_{savings} is observed for smaller values of *cb*. As *cb* increases more jobs are allowed to miss their deadlines. Thus, CRM can operate the CPU at a lower average frequency and achieve higher savings in *E* at higher values of *cb*. This higher saving in energy is accompanied by an increase in the batch completion times (see Figure 7.9). An increase in *cb* increases the deadlines for the jobs in the batch, and a higher batch

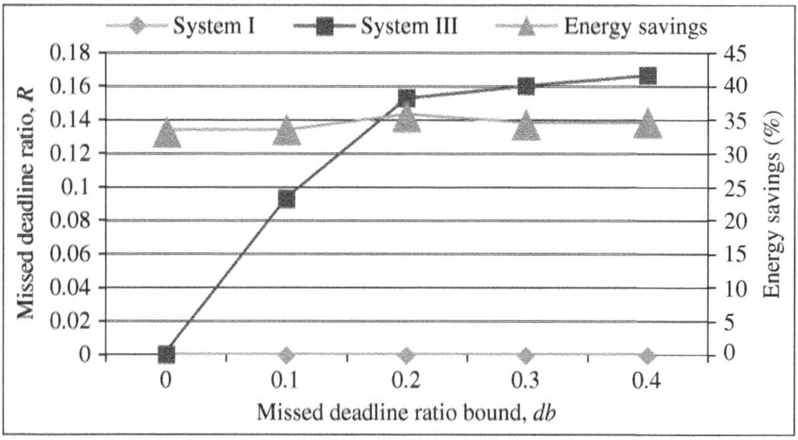

Figure 7.10 Impact of the missed deadline ratio bound. Source: Gregory and Majumdar (2016b)/with permission of IEEE.

completion time is produced. Allowing a later completion time for jobs provides more flexibility in scheduling for the CP optimizer in NEA.

7.5.2 Impact of the Missed Deadline Ratio Bound

The effect of varying the Missed Deadline Ratio Bound on performance is presented in Figure 7.10. For any given value of *db*, System II achieves an energy saving in comparison to System I. As the Missed Deadline Ratio Bound *db* increases CRM is allowed to miss more deadlines. This enables CRM to use a lower average CPU operating frequency. This tends to increase the missed deadline ratio *R* for System II. *R* for System I seems to be independent of the variations in *db*. *This is* because the NEA resource manager's objective does not concern *R* and focuses on minimizing the number of jobs that miss their deadlines. As a result, the number of missed deadlines is observed not to change with *db*.

7.5.3 The Impact of Deadline Multiplier

The impact of varying the deadline multiplier in the workload model is captured in Figure 7.11. As expected, increasing the deadline multiplier increases job deadlines and provides more flexibility in resource management. This allows the resource manager to miss more deadlines and the missed deadline ratio for both System I and System II is observed to increase with an increase in *dm*. A significant energy savings, 35% approximately is observed. The energy savings do not seem to change with *dm*. For a given *dm*, the energy savings achieved by System II are observed to be accompanied by a higher proportion of

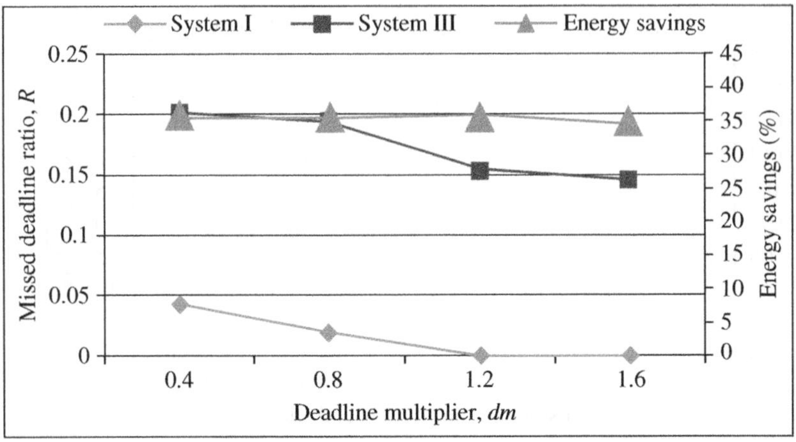

Figure 7.11 Impact of deadline multiplier. Source: Gregory and Majumdar (2016b)/with permission of IEEE.

deadline misses by System II in comparison to System I, which focuses solely on reducing the number of deadlines missed without any concern for server energy consumption.

7.6 Reducing the Number of Active Servers

As mentioned earlier, another approach to energy savings is to reduce the number of active servers when possible. Disabled servers consume a much lower energy in comparison to servers that are active. Thus, such a dynamic reduction in the number of active servers can lead to an overall savings in terms of server energy consumption.

The motivation for this technique is based on the observation that resource management techniques such as EAMR-RM that are based on DVFS are not that useful in reducing energy consumption on lightly loaded servers. This is because lowering CPU energy consumption is not very effective when the energy consumption of the server itself is low. It may be beneficial to transfer tasks from servers with low CPU utilizations to other servers and disable the servers with low CPU utilization. Existing work in the context of various other systems demonstrates the effectiveness of such workload consolidation and is discussed in the next paragraph.

Workload consolidation for reducing overall energy consumption due to lightly loaded servers by disabling unneeded resources is discussed in Kim et al. (2007). This paper presents an energy-efficient VM placement method for batch workloads of MapReduce jobs on clouds. The approach effectively utilizes the trade-off between efficient spatial fitting of VMs on servers to minimize the number of servers required to execute the batch of jobs and balanced temporal fitting of VMs having similar execution times to ensure servers run at high utilization while they remain active.

Reducing energy consumption in clouds subjected to an open stream of deadline-constrained jobs without precedence relationships is considered in (Koomey 2011). The proposed approach advocates putting servers in sleep mode when they are observed to be idle. A threshold-based policy is used to decide when to activate such servers. When the number of jobs waiting in a First Come First Served queue exceeds a specified threshold value of N, sleeping servers are woken up for executing the waiting jobs.

Combining the DVFS-based technique with workload consolidation is a feasible hybrid approach (Gregory 2017). There are two components of energy consumption: the first is the energy consumed by the CPU for executing tasks and the second is the energy consumed by an idle server that is not executing any tasks. The first can be reduced by using DVFS to slow down the CPU so that the slack times in task execution can be traded off for reducing CPU energy consumption. The second component can be reduced by the temporary disabling of idle servers. Both of these need to be performed in such a way that the deadlines of the jobs that are accepted for execution on the system are met.

Once possible way to devise such a combined technique is described next. The resource manager classifies servers into two classes: *active* and *nonactive*. The active servers are characterized by a CPU load that is higher than a given threshold Th. Different metrics that can be used to measure the CPU load include average CPU utilization or the number of waiting tasks on the server. The tasks from the other servers that are characterized by a CPU load that is lower than Th are transferred to one of the active servers and the respective server is deactivated (put in sleep mode). The following set of rules can be used in order to achieve these objectives.

(1) Server Classification: Classify the servers in the system into classes based on load.
(2) Server Deactivation: If the CPU load of a server is below Th transfer jobs/tasks that have not yet started execution to other active servers. Deactivate the server when the last of the remaining tasks completes execution. If such a server cannot be found without missing deadlines of some accepted jobs, then do not perform task transfer.
(3) Selection of Servers: Select an active server that receives the transferred task(s) tasks in such a way that the deadline constraints of tasks are not violated. A CP-based approach as employed by EAMR-RM can be used.
(4) Server activation: When a newly arrived task cannot be allocated with the current number of active severs activate one or more of the deactivated servers.

7.7 Summary

Energy consumed by data centers is an important concern and is a significant component of the data center's operational cost. This chapter focused on resource management techniques that are directed at reducing the energy consumed by the server CPUs when running MapReduce jobs and meeting the deadline specifications for the jobs. Both systems subjected to open streams of MapReduce job arrivals as well as systems for running a batch of MapReduce jobs were considered. Resource management techniques that are based on constraint programming were described for both of these environments.

Two different techniques, one based on DVFS that alters the CPU frequency of servers and the other based on workload consolidation with idle server deactivation, were used. Results of simulation-based performance analysis were discussed to provide insights into the behavior and performance of the resource management techniques.

Exercises

1 Consider the EAMR-RM algorithm. Explain the role of each constraint described in Figure 7.2.

2 Consider the CRM system. Explain the role of each constraint described in Figure 7.8.

3 DVFS and controlling the number of active servers are two different techniques for energy aware resource management. Write an algorithm that can combine these two techniques.

4 How will you extend EAMR-RM to a system with soft deadlines? On such a system a given proportion of the ARs are allowed to miss their deadlines. Discuss how you will extend the EAMR-RM algorithm to handle ARs with soft deadlines when (a) the soft deadline feature can be utilized to increase the maximum arrival rate that can be handled and (b) the soft deadline feature can be utilized to improve energy savings.

5 Discuss the impact of error in user estimation of execution times of Map and Reduce tasks on EAMR-RM.

6 A nonmonotonicity in energy saved is displayed in Figures 7.4 and 7.6 in response to an increase in the arrival rate and in the number of resources, respectively. Discuss the possible rationale for such a behavior captured in the two graphs.

References

A. Gregory, "Energy Aware Resource Management for MapReduce Jobs", M.A.Sc. Thesis, Department of Systems and Computer Engineering, Carleton University, Ottawa, Canada, 2017.

A. Gregory, S. Majumdar, "Energy Aware Resource Management for MapReduce Jobs with Service Level Agreements in Cloud Data Centers", *Proc. 6th IEEE International Symposium on Cloud and Service Computing (SC2)*, Nadi, Fiji, December 2016a.

A. Gregory, S. Majumdar, "A Configurable Energy Aware Resource Management Technique for Optimization of Performance and Energy Consumption on Clouds", *2016 IEEE International Conference on Cloud Computing Technology and Science (CloudCom)*, Luxembourg, Luxembourg, 2016b, pp. 184–192, doi: 10.1109/CloudCom.2016.0041.

J. P. Halimi, B. Pradelle, A. Guermouche, N. Triquenaux, A. Laurent, J. C. Beyler, W. Jalby, "*Reactive DVFS Control for Multicore Processors,*" Green Computing and Communications (GreenCom), *2013 IEEE and Internet of Things (iThings/CPSCom), IEEE International Conference on and IEEE Cyber, Physical and Social Computing*, Beijing, 2013, pp. 102–109.

Hewlett-Packard Corporation, Intel Corporation, Microsoft Corporation, Phoenix Technologies Ltd., and Toshiba Corporation "Advanced Configuration and Power Interface Specification, Revision 5", 6 December 2011.

K.H. Kim, R. Buyya, J. Kim, "Power Aware Scheduling of Bag-of-tasks Applications with Deadline Constraints on DVS-enabled Clusters", *7th IEEE International Symposium on Cluster Computing and the Grid (CCGrid '07)*, Rio De Janeiro, 2007, pp. 541–548.

J.G. Koomey, "*Growth in Data Center Electricity Use 2005 to 2010*", Analytics Press, Oakland, CA, 2011.

C. Piguet, C. Schuster, J.L. Nagel, "Optimizing Architecture Activity and Logic Depth for Static and Dynamic Power Reduction", *Circuits and Systems, 2004. NEWCAS 2004. The 2nd Annual IEEE Northeast Workshop on*, 2004, pp. 41–44.

L. Wang, G. von Laszewski, J. Dayal, F. Wang, "Towards Energy Aware Scheduling for Precedence Constrained Parallel Tasks in a Cluster with DVFS", *Cluster, Cloud and Grid Computing (CCGrid), 2010 10th IEEE/ACM International Conference on*, Melbourne, VIC, 2010, pp. 368–377

8

Streaming Data and Complex Event Processing

8.1 Introduction

Data are an important resource, and effective management of streaming data is important for providing services to users and achieving high system performance. Various types of sources of streaming data are possible. These include sensors in IoT-based systems producing sensor data periodically and Twitter (also known as X) and other social networks generating message streams based on inputs received from their respective users. Examples of IoT-based systems include sensor-based smart cyber–physical systems (Majumdar 2018), such as bridges fitted with sensors for monitoring the structural health of the bridge (Lau et al. 2013), smart houses that use sensors for controlling their temperature and energy usage and sensor-based aerospace engines recording the sensor data from various engine components as well as remote health care systems.

Streaming data is characterized by both the volume and velocity of the streaming workload. One of the challenges of streaming data is the dependency of workload processing and resource management on data content. Two types of streaming systems are described in Section 8.2, which are then discussed in the remaining parts of the chapter. The first part (Sections 8.3–8.5) concerns resource management techniques that can be useful in the context of various different types of stream processing, whereas the second (Sections 8.6–8.7) focuses on a special type of stream processing that concerns the detection of complex events (CEs) from an event stream.

The first part of the chapter discusses two classes of streaming environments. The first class of environments (discussed in Sections 8.3–8.4) is dedicated to a single application, and the second, called multitenant environments (discussed in Section 8.5), allows the execution of multiple applications simultaneously on the same platform. Section 8.3 discusses dynamic scheduling and describes techniques that allow the association of priority with specific data elements so that

Resource Management on Distributed Systems: Principles and Techniques,
First Edition. Shikharesh Majumdar.

high-priority (HP) items can be processed with a shorter delay. Such techniques allow the priority of processing the various data elements to change at run time. The deployment of the scheduling techniques on the well-known Spark platform is considered. An introduction to the streaming components of Spark is presented in Section 8.3. A data-driven priority scheduler (DDPS) that allows the processing of some data elements with a higher-priority and its deployment on Spark is then described in Section 8.4. A performance analysis of the system is also included at the end of the section.

Section 8.5 focuses on multitenant streaming environments in which multiple streaming applications each with its own priority share the resources provided by the same stream processing platform. The resource management problem on Apache Storm, a multitenant streaming platform is presented. The resource management challenges of such a system are described first. Storm's isolation scheduler that does not support the notion of priorities as well as a static and dynamic priority scheduler (DPS) that can mitigate the shortcomings of the isolation scheduler are described. For the static scheduler, the priorities are fixed and do not change with time, whereas the dynamic scheduler allows the priorities to vary with time. A performance analysis of these techniques is also included.

The second part of the chapter discusses CE processing (CEP). Section 8.6 introduces the concept of CEs, each of which is often the result of the simultaneous occurrence of multiple raw events (REs). CEP is used in big data analytics for analyzing multiple streams of data in real time to generate alerts or to indicate possible opportunities to be exploited on the system. CEP and platforms to run such CEP applications are discussed. An example system that performs remote patient monitoring (RPM) using CEP on a mobile phone is described in Section 8.7. Section 8.8 summarizes the key issues discussed in the chapter.

8.2 Management of Streaming Data

A stream comprises a set of data tuples. A sequence of such data tuples is often called a message. This chapter focuses on the management of such streaming data, and two examples of stream processing are provided in this section.

The case for systems that require dynamic priority-based scheduling is presented. Such systems are used for processing messages in which certain messages are delay sensitive and thus need to be processed ahead of other lower-priority messages. The default Spark scheduler used in the processing of streaming data (see Section 8.4.4) executes jobs in the same order in which they arrived. It is thus unable to process higher-priority messages before the lower-priority messages that arrived in the system earlier. A system may, however, need to execute such a message earlier based on its content which indicates the higher priority of the message. A priority scheduler that can schedule jobs on a system in which the priority of the job depends on the content of the message is thus crucial for resource management on these systems. Techniques for scheduling messages, the priority of which can change dynamically with time are discussed in Section 8.3.

In streaming systems, especially sensor-based systems, events, also referred to as raw events (REs), are said to be generated when a specific sensor output crosses a given threshold value. If the output of a heart rate sensor attached to the body of the user, for example, exceeds a specific value, a RE called "High Rate" is said to have been generated. The simultaneous occurrence of multiple REs gives rise to a *CE*. CEs may be of special interest in systems that attempt to detect an impending problem before it occurs. For example, a CE generated on an RPM system (discussed in Section 8.7) may indicate an impending problem for a patient with chronic heart disease. CEP systems that are subject to such event streams are discussed in Section 8.6.

8.3 Dynamic Priority-Based Scheduling

Most existing approaches to scheduling data elements arriving as part of streaming data do not allow the association of priority with specific data elements so that IIP items can be processed with a shorter delay. Instead, they seem to be focusing on attempting to minimize the average processing delays for all elements. Delays in the processing of data items often result from variations in system load. Existing approaches to handling delay-sensitive data items in the context of varying system loads include (i) adaptive scheduling, (ii) load shedding, and (iii) changing the number of jobs that are processed in parallel based on the current system load. Adaptive scheduling (Das et al. 2014) controls the batch duration, exploiting the trade-off between latency and throughput. For a small batch duration, the batch-processing latency is reduced, and the system throughput is increased. So, on systems characterized by varying loads, the system reduces the batch duration (thereby decreasing processing latency) when the load intensity is low and increases it when the load increases so that it can handle the higher input message rate. With such a system, scalability is achieved at the cost of increased processing latency.

The second technique called DSlash (Birke et al. 2017) is a data controller that processes data as fast as possible based on a given target data processing latency. It performs load shedding by discarding data items of an age higher than a given threshold. One of the disadvantages of this technique is that some of the data items are lost.

The third technique is based on A-Scheduler (Cheng et al. 2017), which dynamically modifies the number of jobs that are processed in parallel. By running jobs in parallel, both latency and throughput can be improved. The shortcoming of this approach is that static resource provisioning can improve system performance only when the system is underutilized. When the resources are fully utilized, the degree of parallelism in job execution cannot be increased because of the lack of idle resources, and thus, system performance cannot be improved further.

DDPS discussed in Section 8.4 alleviates a number of the shortcomings of the three techniques discussed in the previous paragraphs. In particular, it enables the association of priority with data items that enable shortening the processing latency for HP data items that are not handled by any of the approaches (Birke et al. 2017; Cheng et al. 2017; Das et al. 2014) discussed earlier.

8.3.1 The Spark Streaming Data Processing Platform

High-performance big data platforms are often deployed in the analysis of streaming data. A few examples are discussed next.

Apache Spark: is also discussed in Chapter 9. A part of the following discussion includes topics described in the chapter that are augmented by a discussion of Spark Streaming, which is required for the effective handling of streaming data. Apache Spark is a parallel processing framework that is useful for processing large datasets. It addresses a performance shortcoming of its predecessor Apache Hadoop in which the intermediate results of an operation performed on a MapReduce program, for example, needs to be written onto electro-mechanical devices such as disks thus slowing down the system. Apache Spark alleviates this problem by using only in-memory computations thus avoiding the use of disk drives for transferring data between the tasks. The application areas for Spark include e-commerce, social media, entertainment, and health care. Examples of companies that use Spark, such as eBay and Netflix, are described in (DeZyre n.d.).

Spark uses (Apache Spark n.d.-a) a Master–Slave architecture comprising one master and several worker nodes. When Spark is deployed on a cluster, the nodes receive resources from a cluster manager.

An application is launched in the master node of Spark. The master node is referred to as the driver and communicates with the worker nodes, which are also known as executors. The driver program creates SparkContext, and is responsible for creating resilient distributed datasets (RDDs) and executing all transformations and actions on them (Apache Spark n.d.-b). A transformation is a function that returns a new RDD by modifying existing RDD(s). An RDD is a read-only data structure that is memory resident and is used to represent large datasets. RDDs are distributed collections of objects. They are used in conjunction with two types of operations: *transformations* and *actions*. A transformation creates a new RDD based on an existing RDD but does not return any value. An action, on the other hand, returns a value without creating a new RDD. An example of a transformation is "filter" that returns a new RDD based on the filter condition (Apache Spark n.d.-b). An example of a spark action is "collect," which returns the contents of an RDD to the driver node in the Spark cluster (Apache Spark n.d.-b).

8.3.1.1 The Spark Streaming System

Spark Streaming (Apache Spark n.d.-c) is an extension of Spark Core used for stream processing. It is both scalable and fault tolerant. It can process data generated by various stream sources in real time. Examples of such data include sensor outputs in IoT-based systems, system logs, and information updates posted by users on social media.

The stream of data is divided into small batches. These batches are built on RDDs. Data sources may be directly connected to the Spark system through a TCP socket for example or data may arrive through a publish/subscribe system such as Kafka (Apache Software Foundation n.d.). Data is sent by sources to Kafka whereas it gets stored under specific Kafka topics depending on the type of the data. It is then partitioned and processed by the Spark Streaming component.

Spark Streaming is a fault-tolerant microbatching engine that is often used for real-time streaming applications. The output of Spark Streaming can be sent to databases as well as file systems. This system is used to apply transformations to data items in streams. These transformations are similar to the ones that are applied to RDDs. An example is the "map" transformation that applies a given function to each data item in the stream.

Streams are represented as discretized streams referred to as Dstreams each of which contains RDDs that are data items from a specific time interval. These are depicted on the horizontal axis of Figure 8.1. Any transformation performed on

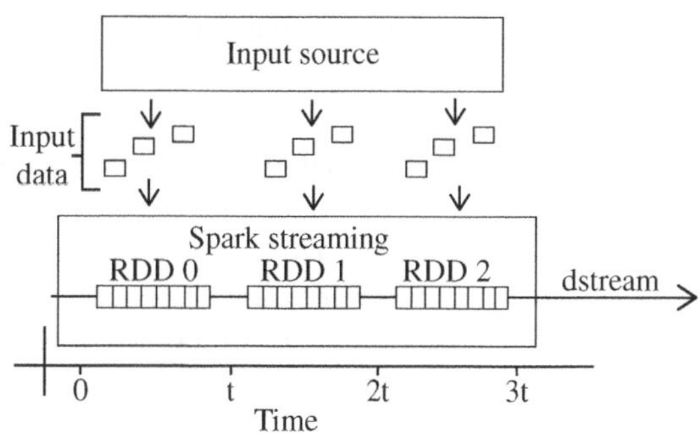

Figure 8.1 RDDs in the spark streaming system. Source: Ajila and Majumdar (2019)/with permission of IEEE.

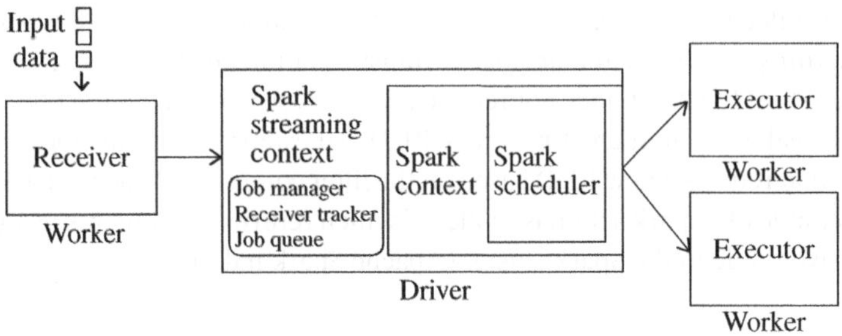

Figure 8.2 Components of the spark streaming system. Source: Ajila and Majumdar (2019)/with permission of IEEE.

the D_{stream} is applied to each item in the stream. The output of Spark Streaming is a series of output batches.

The main components of Spark Streaming are captured in Figure 8.2. The receiver runs on a Worker node and looks for socket connections. Once such a connection is established, it receives data items from the socket. A receiver tracker that tracks them stores them into an RDD prior to the start of a batch interval. At the start of a batch interval, the job scheduler creates jobs for the RDDs. A job corresponds to the collection of these RDDs created from the input data items received within a specific time window and the transformations that are applied to it. Each job is then placed in a job queue that is processed in a FIFO manner.

One of the problems with using a FIFO queue is that priorities cannot be associated with data items. Associating such a priority with data items and performing data priority-driven scheduling is crucial in certain systems. Consider, for example, a system in which real-time analysis of tweets is performed by a police department. A tweet that contains the word “explosion” may indicate a security threat and needs to be analyzed immediately. A FIFO queue does not allow such out-of-turn scheduling. A data priority-driven scheduling technique is described in Ajila and Majumdar (2019), which allows the association of dynamic priority with data items and the analysis of an HP tweet before the analysis of other tweets that have arrived in the system earlier.

8.4 Data-Driven Priority Scheduler (DDPS)

The DDPS technique described in Ajila and Majumdar (2019) supports two priority levels: HP and low priority (LP). The authors of the article claim that it can be extended to handle more than two priority levels with a modest effort. The user needs to perform the classification of input data items first. The priority of these items arriving on the system as part of messages are then determined based on the respective classes. For example, "explosion" and "bomb" can be classified as data items in the HP class.

If a message contains a single HP keyword the priority of the message is elevated. If a message contains *n* HP keywords its priority becomes *n* times the priority of a message containing one HP keyword. The message with a higher-priority is processed prior to a message with a lower priority.

The data-driven priority scheduler comprises three algorithms: Algorithm 1, Algorithm 2, and Algorithm 3. Algorithm 1 is responsible for receiving the streaming data items and placing them in Memory. Algorithm 2 takes these data items and, creates jobs form these items and assigns job priorities based on the user specifications. Algorithm 3 is used for sorting the elements in the job queue. Brief descriptions of the algorithms are provided next.

8.4.1 Algorithm 1

Algorithm 1 is explained with the help of Figure 8.3.

The algorithm runs continuously. It reads the data items from a socket (lines 3–8) and stores them in memory (line 9). Algorithm 1 runs concurrently with Algorithm 2, which is explained in the following subsection.

Figure 8.3 Algorithm 1: High-level description. Source: Ajila and Majumdar (2019)/with permission of IEEE.

Algorithm 1: Algorithm for receiver

```
//The following algorithm is run by the receiver
//Input:
// UBList unattached blocks list. Memory blocks that
// have are not been assigned to a RDD
1: Begin
2: while True
// receive data from socket connection
3:    D_in ← receiveData()
4:    if full(MB_current)
//if the current memory block is full, create a new one
//and add it to the unattached memory block list.
5:        MB_new ← allocateNewBlock()
6:        insert MB_current in UBList
// Use the new block from now on
7:        MB_current ← MB_new
8:    endif
9:    insert D_in in MB_i
10: endwhile
11: End
```

8.4.2 Algorithm 2

Algorithm 2 is explained with the help of Figure 8.4. At the beginning of each batch interval, the data from the memory blocks (MBs) inserted by Algorithm 1 are added to an RDD (lines 4–7). This is accomplished by removing each memory block MB_i from the UBList and putting it in the RDD. The algorithm goes through the RDD and tries to identify HP keywords in each element D_i in the RDD. The priority of the respective job is then computed based on the number of HP keywords (listed in HPList detected in the RDD [lines 8–14]). The RDD is then assigned to the new job. The job is responsible for the RDD and all the transformations that are applied to the RDD. The job is then passed on to the job manager (line 16).

The job manager is responsible for managing the job queue by using Algorithm 3 that is explained in the following subsection.

Algorithm 2: Algorithm for creating jobs

```
// Algorithm is run by receiver tracker
// Input:
// UBList unattached blocks list.
// HPList a list of all high priority keywords
// Output:
// Job
1:  Begin
2:  while True
3:      create RDD and Job
4:      for each MB in UBList
// take all memory blocks in the unattached list and add
to
// the RDD
5:          assign MB_i to RDD
6:          remove MB_i from UBList
7:      endfor
// scan RDD for high priority keywords
8:      for each Keyword in HPList
9:          for each D_i in RDD
10:             if D_i contains Keyword
// if high priority keyword is found increase the priority
of the Job
11:                 increasePriority (Job)
12:             endif
13:         endfor
14:     endfor
15:     assign RDD to Job
16:     send Job to JobManager
17:     wait BATCH_INTERVAL
18: endwhile
19: End
```

Figure 8.4 Algorithm 2: High-level description. Source: Ajila and Majumdar (2019)/with permission of IEEE.

8.4.3 Algorithm 3

This algorithm is explained with the help of Figure 8.5.

This algorithm goes through the job queue and ensures that the queue remains sorted in order of job priority. The higher the job priority, the higher its position in the queue. Each job in the queue is processed starting with the highest priority item in the queue. If the priority of the incoming job is higher than that of the highest priority job in the queue, then the incoming job is placed in front of this job (lines 3–4). If the two priorities are equal, the tie is broken by using the age of the job (lines 7–8). If the priority of the job is smaller, then its priority is compared with that of the next job in the queue, and so on.

8.4.4 Performance Analysis

The performance analysis of the techniques was performed through a prototype implementation deployed on Apache Spark. The system is explained with the help of Figure 8.6. A number of modifications were performed to the existing Spark Streaming system for implementing the prototype. The priorities of the keywords are specified by the user before running the application and are stored in a buffer.

Figure 8.5 Algorithm 3: High-level description. Source: Ajila and Majumdar (2019)/ with permission of IEEE.

Algorithm 3: Algorithm for sorting jobs in the job queue

```
// The following algorithm is run each time the job
// manager receives a job
// Input:
// Incoming job J_in
// Priority job queue Q = {J_1 ... J_n}
1: Begin
// loop through all the jobs in the priority queue
2: for each J_i in Q
// check if the incoming job has a higher priority than the
// current job in the queue
3:    if priority(J_in) > priority(J_i)
4:       insert J_in in front of J_i in Q
5:       break forloop
// when handling jobs with equal priorities, use age of a
job to break the tie
6:    else if priority(J_in) == priority(J_i)
7:       if age(J_in) > age(J_i)
8:          insert J_in in front of J_i in Q
9:          break forloop
10:      endif
11:   endif
// Check the next element in the queue
12: endfor
13: End
```

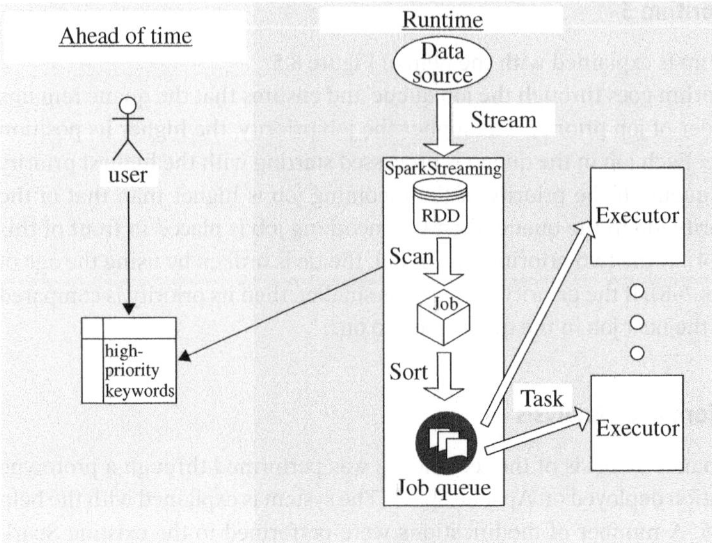

Figure 8.6 Description of the proof-of-concept prototype. Source: Ajila and Majumdar (2019)/with permission of IEEE.

This list of keywords is read during run time by the modified receiver tracker that also scans the incoming RDDs for the existence of HP keywords. The modified job manager stores the jobs in the priority queue for jobs. The jobs are executed by tasks running on the executors.

The impact of the arrival rate of messages on the system (λ) on the average latency is presented in Figure 8.7, which compares the performance of DDPS with first-in first-out (FIFO) the default scheduling policy of Spark (Apache Spark n.d.-d). The end-to-end latency of a message is the difference in time between the time when the processing of a message is completed and the time at which it was sent. The average of the end-to-end latencies for all the messages is the average latency presented in Figure 8.7. There are four lines in the graph: one corresponding to the HP messages (labeled HP), one to LP messages (labeled LP) observed with DDPS, one for HP messages (labeled HP–FIFO), and one for LP messages (labeled LP–FIFO) observed with FIFO. As λ increases, contention on the system increases, thus increasing the average latency for the messages irrespective of the priority of the messages. The impact of priority-based scheduling becomes visible at higher values of the arrival rate. At such high arrival rates, DDPS leads to the smallest average latency for HP messages (see line labeled HP). This benefit comes, however, at the cost of LP messages experiencing the highest average latency. For FIFO, both HP and LP messages lead to comparable average latencies as demonstrated by the overlapping of the lines HP–FIFO and LP–FIFO.

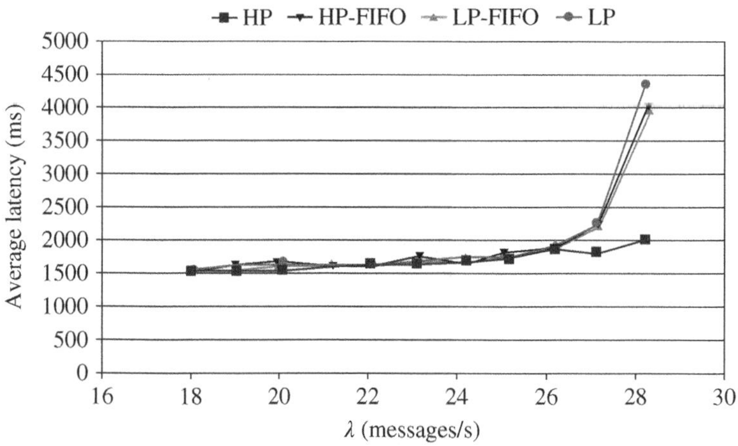

Figure 8.7 Performance of scheduling policies. Source: Ajila and Majumdar (2019)/with permission of IEEE.

8.5 Multitennant Systems

Resource management is an important problem in the context of multitenant streaming systems in which multiple streaming applications share the resources provided by the same string processing platform. In such systems, resources need to be portioned among multiple applications. One possible approach is to divide the resource statically among the competing applications so that each application executes without interference from the other applications running on the system at the same time. The number of resources dedicated to an application can be based on information about application resource requirements provided by the programmer. The problem is relatively simple when there are enough resources on the system to meet the demands of all the applications. Careful resource allocation and scheduling is required, however, on a resource constrained system on which the aggregate resource demands of all applications is larger than the size of the available resource pool.

This resource management problem and its solutions are described in the context of storm a well-known stream processing system that is discussed next (Apache Storm n.d.-b).

8.5.1 Apache Storm

The storm processing platform provides users with APIs for developing stream processing applications. Such applications are referred to as *topologies*. Each topology is subjected to a stream comprising an unbounded number of data tuples.

Spouts and bolts are the important components of a topology. Each of these is described.

Spouts: are used to connect to various types of message sources such as message brokers including Kafka. A spout contains the application logic to connect with the message brokers and push the respective data stream to the other processing elements.

Bolts: are the processing components of the topology and contain the user-defined logic for a given application. A bolt can perform various operations, such as filter and aggregate on the tuples in a received stream.

A storm (see Figure 8.8) cluster is based on a master–slave architecture. The cluster comprises two types of nodes: a single Nimbus node and one or more Supervisor nodes.

Nimbus Node: This is the master node in a storm cluster where storm topologies are submitted by the application developers. Nimbus runs a Nimbus daemon that distributes the tasks in the storm applications/jobs to various worker nodes using a system referred to as the scheduler. It also monitors the cluster state. In case of a failure of any Supervisor nodes, it tries to restart the failed node or move its tasks to the other supervisor nodes in the cluster.

Supervisor Nodes: There can be one or more supervisor nodes running its own supervisor daemon. Each supervisor node runs a predetermined number of worker processes. Each worker process can run one or more components of a storm topology. The worker processes are started/stopped by the respective supervisor daemon.

In addition, a storm cluster contains zookeeper nodes (Chakraborty and Majumdar 2017) for providing synchronization and state management in the cluster.

Worker Processes: Each worker process is typically a Java Virtual Machine (JVM) that runs a subset of the components of a given topology. Each worker process runs in a specific slot in the respective supervisor node.

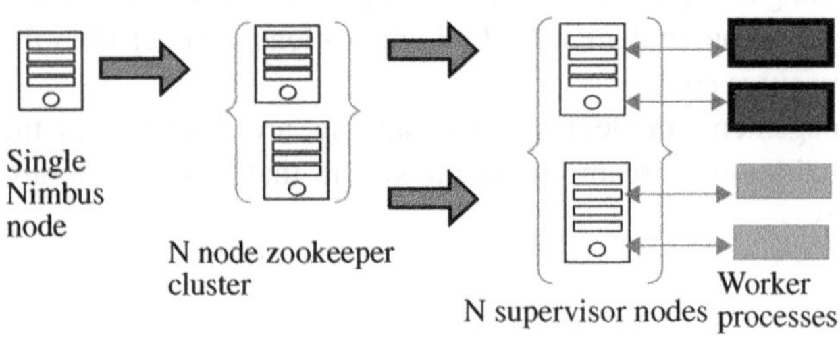

Figure 8.8 A storm cluster with *N* supervisor nodes. Source: Chakraborty and Majumdar (2017)/Association for Computing Machinery.

Executors: are threads spawned by a given worker process. Each executor runs one or more tasks defined by the user.

Task: A task is a running instance of a storm component: spout or bolt. Processing of instructions is performed by a task that runs within the corresponding executor.

A *parallelism hint specified* by the topology developers is the number of concurrent executors for each component of the topology that would run inside the worker processes of the supervisor nodes (Chakraborty and Majumdar 2017). The executors running in the worker processes of supervisor nodes in a storm topology can be modeled by a directed acyclic graph.

8.5.2 Resource Management on a Multitenant Storm Cluster

When a topology is submitted by a user, the Nimbus node invokes the Nimbus scheduler for mapping all the executors in the topology to the available worker processes of the various supervisor nodes. If this is not possible and a topology does not receive any resource, the respective topology will be resource-starved and will not be able to perform any processing. If allocation is performed for some of the components of a given topology, then this may hinder the performance and functional correctness of the respective application.

When the total number of supervisor nodes in a storm cluster is smaller than the sum of the number of supervisor nodes required by all the topologies in the cluster, the cluster is said to be resource constrained. The user of a topology specifies the number of supervisor nodes and the number of executor processes required for running the topology. After reading this information from the configuration file, the scheduler allocates the specified number of supervisor nodes and executor processes to the topology. A Round Robin strategy is used for distributing the worker processes among supervisor nodes.

8.5.2.1 Isolation Scheduler

Storm provides the isolation scheduler (Apache Storm n.d.-a) that makes it easy to share the cluster resources among multiple topologies. The isolation of a topology refers to the running of the topology without interference from the other topologies executing at the same time. Each topology has a dedicated set of resources that avoids interference from other topologies during the running of a given topology. The scheduler uses the information on resource requirements submitted by the user when submitting the topology to the storm system. Although the isolation scheduler provides the desired isolation on a multitenant storm cluster it can lead to starvation of certain topologies when the cluster is resource constrained and the total number of requested resources from all topologies is larger than the total number of available resources on the cluster. Isolation Scheduler allocates resources to topologies in a First Come First Served manner.

Thus, for some topologies that are considered later by the scheduler there may not be enough remaining resources to satisfy their requirements. Another shortcoming of the isolation scheduler is that it cannot handle priorities associated with topologies. The two schedulers described next are aimed at addressing these issues.

8.5.2.2 Static Priority-Based Scheduler (SPS)

Topologies are associated with priorities: a lower-priority number implies a higher-priority. The priority of a topology is static and cannot change with time. The system comprises a Priority Manager (PM) and SPS. While submitting a topology the user needs to specify a priority value and the desired minimum number of supervisor nodes to PM. The algorithm for scheduler, SPS, is captured in Figure 8.9.

There are two parameters related to the degree of parallelism for a given topology. The *desired number of supervisor nodes* and the *minimum number of supervisor nodes* specified by the topology user are different from one another. The first specifies the number of supervisor nodes required for maximizing the degree of parallelism for the topology whereas the second is the bare minimum number of supervisor nodes required by the topology such that system performance does not fall below the acceptable level. To prevent topology starvation, the total number of supervisor nodes in the cluster must be more than or equal to the sum of the minimum number of supervisor nodes for all the topologies sharing the cluster.

The SPS algorithm is described in Figure 8.9 while the terminologies used in its description are introduced in Figure 8.10.

The SPS algorithm (see Figure 8.9) obtains the total number of supervisor nodes (C) in the cluster (line 1). Note that this algorithm is run only when there is enough number of supervisor nodes in the cluster such that topology starvation does not occur. If the sum of the minimum number of supervisor nodes for all topologies to be run on the cluster is equal to C, then the minimum number of supervisor nodes required by the topology is allocated to the respective topology [lines 5, 6]. Otherwise, the supervisor nodes are distributed proportionally among the topologies using the priority of the topologies. This is accomplished in lines 10–23. ST, the sum of all sd_{ij} values is computed in line 10. If ST is lower than C, the total number of supervisor nodes to be distributed, X, is capped at C (line 11). The supervisor node allocation for each priority level j is computed in line 12 to line 22. The total number of supervisor nodes to be allocated to the topologies at the jth priority level is computed as PF_j (line 14). In SPS, a higher proportion of the available resources are allocated to higher-priority topologies. This is accomplished for each topology

Figure 8.9 The SPS algorithm. Source: Chakraborty and Majumdar (2017)/Association for Computing Machinery.

1 *Get* C from the cluster

2 $if\ C == \sum_{j=1}^{p} \sum_{i=1}^{i_j} sm_{ij}$

3 **for** *each* $j \in$ *in PR*

4 **for** *each* tp_{ij} *in* $\in TP_j$

5 $cn_{ij} = sm_{ij}$

6 add cn_{ij} to set CN

7 *end for*

8 *end for*

9 *else*

10 $ST = \sum_{j=1}^{p} \sum_{i=1}^{i_j} sd_{ij}$

11 X=min(C,ST)

12 **for** *each* $j \in PR$

13 $SP_J = \sum_{i=1}^{i_j} sd_{ij}$

14 $PF_j = \min(X, round(C * \left(\frac{SP_j}{ST}\right)))$

15 $Y = PF_j$

16 *for each* $tp_{ij} \in TP_j$

17 $cn_{ij} = \min(Y, round(PF_j * \left(\frac{sd_{ij}}{SP_j}\right)))$

18 add cn_{ij} to set CN

19 $Y = (Y - cn_{ij})$

20 *end for*

21 $X = (X - PF_j)$

22 *end for*

23 *end else*

24 **call** *ResourceAllocator()*

tp_{ij} in the jth priority level in PR where the number of supervisor nodes allocated to topology tp_{ij} is computed as cn_{ij} that is added to the set CN (lines 16–18). This is performed for each priority level and the resource Allocator function is called with CN as the argument so that the computed number of supervisor nodes for each topology are allocated to the respective topology (line 24).

8.5.2.3 Dynamic Priority Scheduler

DPS uses dynamic priority indications from topologies at run time to perform resource allocation to strategies. It supports two priority levels: high and low. As in the case of SPS, the system comprises two components: PM and the scheduler

Symbol	Description
$\text{PR} = \{1, 2, 3\ldots p\}$	Set of distinct priority levels; p is the total number of priority levels available. A lower numeric value indicates higher priority
$l_j \ (j = 1, 2\ldots p)$	Total number of topologies in jth priority level.
$\text{TP}_j = \{\text{tp}_{ij}\} \ (j = 1, 2\ldots.p; i = 1, 2\ldots l_j)$	Set of all the topologies in the jth priority level; tp_{ij} is the ith topology of the jth priority level. Note that the total number of topologies submitted to the cluster is $\sum_{j=1}^{p} l_j$
$sd_{ij} \ (j = 1, 2\ldots p; i = 1, 2\ldots l_j)$	Desired number of supervisor nodes for the ith topology in the jth priority level
C	Total number of supervisor nodes available to the cluster
$W_{ij} \ (j = 1, 2\ldots p; i = 1, 2\ldots l_j)$	Total number of worker process desired by the ith topology of the jth priority level. Each slot is used for running separate worker process
$\text{CA} = \{\text{ca}_{ij}\} \ (j = 1, 2\ldots p; i = 1, 2\ldots l_j)$	Set of previously allocated number of supervisor nodes for each topology; ca_{ij} is the number of supervisor nodes already allocated to the ith topology of the jth priority level. A newly submitted topology has a default value 0.
$\text{CN} = \{\text{cn}_{ij}\} \ (j = 1, 2\ldots p; i = 1, 2\ldots l_j)$	Set of computed number of supervisor nodes for each topology; cn_{ij} is the maximum number of supervisor nodes computed by the algorithm for allocation to the ith topology of the jth priority level
$\text{SM} = \{\text{sm}_{ij}\} \ (j = 1, 2\ldots p; i = 1, 2\ldots l_j)$	Set of minimum number of supervisor node requirements of the topologies; sm_{ij} is the minimum number of supervisor nodes required by the ith topology of the jth priority level
$E_{ij} \ (j = 1, 2\ldots p; i = 1, 2\ldots l_j)$	Set of executors for the ith topology of the jth priority level
ST	Total number of supervisor nodes desired by all the topologies submitted
$\text{SP}_j \ (j = 1, 2\ldots p)$	Total number of supervisor nodes desired by all the topologies in the jth priority level
PF_J	the maximum number of supervisor nodes that can be allotted to the topologies of the jth priority level

Figure 8.10 Terminology used in priority-based algorithms. Source: Chakraborty and Majumdar (2017)/Association for Computing Machinery.

Figure 8.11 The DPS algorithm. Source: Chakraborty and Majumdar (2017)/Association for Computing Machinery.

1 *Get C* from the cluster
2 $ST = \sum_{j=1}^{p} \sum_{i=1}^{i_j} sd_{ij}$
3 X=min(C, ST)
4 ***for** each j* $\in$ *PR*
5 $SP_j = \sum_{i=1}^{i_j} sd_{ij}$
6 $P = \min(X, SP_j)$
7 *for each* $tp_{ij} \in TP_j$
8 $cn_{ij} = \min(X, round(P * \left(\frac{sd_{ij}}{SP_j}\right)))$
9 add cn_{ij} to set CN
10 $X = (X - cn_{ij})$
11 *end for*
12 *end for*
13 *end else*
14 ***call** ResourceAllocator()*

DPS. At the beginning, PM assigns the same default priority value to all the topologies. The priority of a topology changes at run time when the topology sends a "trigger ON" signal to PM. The detection of the trigger condition and the subsequent logic is business case dependent and is thus the responsibility of the topology designer. Upon receiving this trigger ON indication, DPS generates a new allocation in which the higher-priority topologies receive a higher proportion of resources. The DPS algorithm is presented in Figure 8.11. After getting the total number of supervisor nodes in the cluster (line 1), X the number of nodes available for allocation is computed as the minimum of C and ST, the sum of the total number of desired supervisor nodes of all the topologies (line 3). DPS then computes the number of supervisor nodes required at each priority level j (iterative loop from lines 4 to 12). The total number of supervisor nodes to be used at priority level j is computed and stored in SP_j. The number of supervisor nodes to be allocated to the topology tp_{ij} is computed from X and the ratio between sd_{ij} and SP_j (line 8). At the end (line 14), the algorithm uses a call to ResourceAllocator to allocate cn_{ij} nodes to each topology tp_{ij}.

8.5.2.4 Performance Analysis

The results of experiments run on a prototype for the SPS and DPS algorithms are described in Chakraborty and Majumdar (2017). The prototype was deployed on an Amazon EC2-based Storm cluster comprising 26 EC2 nodes. A synthetic workload using four-storm topologies was used to ingest the batches of tuples in the data stream from an Apache Kafka system. Further description of the experiments

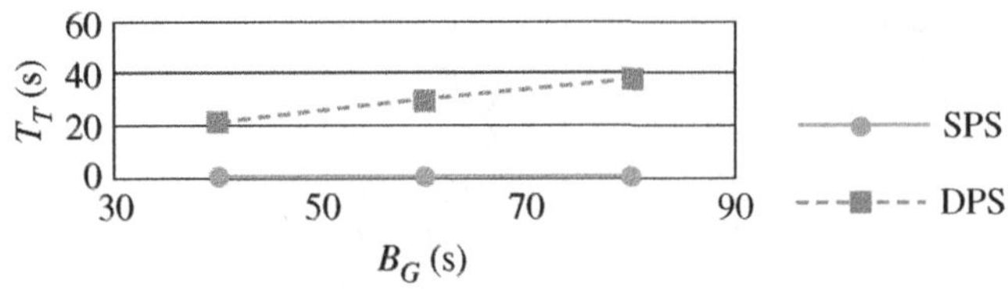

Figure 8.12 Effect of B_G on T_T lower-priority topologies). Source: Chakraborty and Majumdar (2017)/Association for Computing Machinery.

including the details of the system configuration and generation of the workload are not important in the context of this discussion. The interested reader can find the details in Chakraborty and Majumdar (2017).

Figure 8.12 presents the comparison of the two scheduling algorithms, SPS and DPS, when the Batch Gap (B_G) is varied. B_G is the time difference between the two successive batches of tuples generated. The performance metric used is the average tuple latency (T_T) for lower-priority topologies that is the average time taken by the lower-priority topologies to process tuples. For any value of B_G, SPS gives rise to a shorter T_T for the lower-priority topologies. As B_G is increased the performance difference between SPS and DPS seem to increase. The rationale for such behavior is provided next. Note that with DPS, when a trigger ON condition occurs the higher-priority topologies receive a higher proportion of resources, thus starving out the lower-priority topologies. Note that this starvation is temporary and goes away as soon as the trigger indication is switched off. As B_G increases, the time difference between the trigger ON and trigger OFF conditions tends to increase. As a result, the starvation periods for the lower-priority topologies experienced with DPS increases, and the performance difference between SPS and DPS increases. T_T for higher-priority topologies seem to be insensitive to the variation in B_G (Chakraborty and Majumdar 2017).

8.6 Complex Event Processing

A CE is often the result of the simultaneous occurrence of multiple REs. CEP concerns real-time issues in the context of big data analytics. As indicated earlier in the chapter it is often used in analyzing multiple streams of data in real time to generate alerts or to indicate possible opportunities to be exploited on the system. In a sensor-based system, this may be achieved by fusing multiple data streams from multiple sensors inside a CEP engine to match patterns specified by using a Continuous Query Language (CQL) (Luckham 2001) for example. A number of terminologies used in the context of CEP are presented.

Data Stream: A data stream contains an unbounded number of events that arrive continuously. Examples of sources of such data streams include sensors, GPS devices, networks creating packets, tweeter, and devices generating click streams.

Events: An event can be represented by a tuple data type or object in many programming languages. The events in a data stream represent the attributes of the system included in the event. Examples of such attributes are event generation time stamp, the sensor value while monitoring a given phenomenon of interest, the sensor name and a tuple id. Three types of events are described next.

Raw Events (RE): are events generated by a source such as a sensor.

Complex Events (CE): are generated by a CEP engine. CEs typically result from a pattern match observed on streams of lower-level events.

Intermediate Events: are the result of the processing of REs by a CEP engine.

Each RE stream processed leads to the generation of a specific IE stream. Multiple IE streams after processing by the CEP engine can lead to a CE stream.

Dataset: A dataset consists of a bounded number of events that are typically stored in a database.

8.6.1 CQL Query and CEP

A CQL query processes both RE streams and IE streams to produce a CE stream (see example in Figure 8.13). The two RE streams in the figure are received by a selection operator (π) and a projection operator (α) (Dhillon et al. 2018). The selection operator selects predicates based on a number of given selection conditions. The projection operator selects all the stream elements. The two IE streams resulting from the processing of the REs by the respective operators are then processed by a join operator that joins the two streams based on a given key.

CEP is typically performed with the help of a Data Stream Management System (DSMS). As opposed to a Database Management System (DBMS) that performs operations on a finite set of persisted data a DSMS processes unbounded data streams that continuously arrive on the system. A CQL is often used in a DSMS. The CQL queries are static while the data they work on are dynamic.

Figure 8.13 Example of stream processing.

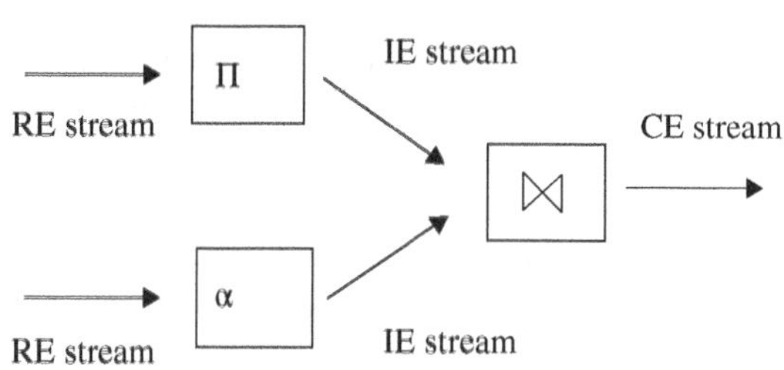

As streams are continuously arriving a one-time processing model is followed by a CEP engine.

The Aurora system is one of the first string processing systems that has DSMS capabilities (Abadi et al. 2003). The builders of Aurora have also provided an imperative CQL called Stream Query Algebra (SQuAl). Further development of Aurora was achieved through the Medusa system (Zdonik et al. 2003) that enabled the scaling of the system over multiple nodes (Rizvi 2005).

Stream processing engines used in Apache Storm, for example, use imperative programming and the user needs to write the logic for performing operations. This is achieved through a Storm topology comprising bolts and spouts. Bolts contain the logic for the task provided by the user, whereas spouts ingest events and are often placed between bolts (see Chakraborty and Majumdar 2017, for example). In CEP, a declarative programming approach is followed for which the programmer does not need to specify the logic for the operations to be performed. Instead, a CEP query, which comprises CEP operators is used for specifying "what" operations (and not "how") need to be performed.

8.6.2 CEP Architecture

CEP engines tend to work in a centralized manner because the various CEP operators share the same state information. Such engines typically have a low latency associated with their operations. They are thus used in various real-time applications.

A discussion of a high-level CEP architecture is presented. The various components of the architecture are briefly described.

Event Source: An event source receives the RE streams from external sources such as sensor outputs. Examples of event sources include Message Queueing Telemetry Transept (MQTT) (Hunkeler et al. 2008), TCP sockets, and Kafka (Apache Software Foundation n.d.).

Event Sinks: An event sink publishes the results generated by the CEP engine. Examples include emails, TCP sockets, Kafka, and dashboards.

Mappers: perform data conversions. There are two types of mappers each of which is described next.

- Source Mapper: performs the format conversion from the custom format used by the input data streams generated by a source to the native format that is understood by the CEP engine. Examples of formats used by events include XML, text, comma-separated values (CSV), Java Script Object Notation (JSON), WSO_2 (discussed in Section 8.7) event, and text.
- Sink Mappers: convert the native data type for the events generated by the CEP engine to the event type required by the respective event sink.

Blocking Queues: These are used in a CEP system for temporarily storing events before sending them to the CEP engine. Since there are both producers and consumers of data in the system providing thread safety when multiple producers and consumers are trying to access the queue elements at the same time is important. Blocking queues provide such thread safety on the system.

In-Memory Tables: The persistence of events can also be achieved with in-memory tables. The events in a data stream are typically converted into a relationship before persisting them in a database.

Callbacks: A callback is used to receive the CEs from a CEP engine. Two important types of callbacks are described.

- System Callback: is used to receive CEs for a specific stream. A CEP engine can generate multiple CEP streams. A system call back is used to receive CEs from a specific stream by providing a string identifier. Whenever a CE is detected the stream callback receives this event.
- Query Callbacks: receive all the data streams generated by a CQL query. These include both IEs and CEs.

CEP Operators: are used in CQL. Examples include select and join. There are arithmetic, logical, and comparison operators. An operator in each of these categories can be blocking or nonblocking.

Blocking operators require all the events in the data stream. These operators save the states of pervious tuples for producing the result. For example, the max operator needs the state information of all data stream events before computing the maximum value. Aggregate, Join and Cartesian product are additional examples of such operators. A nonblocking operator on the hand generates the results as and when events occur on the system. Examples include conjunction and disjunction operators.

Windows: Data streams are continuous. Windows are used to select a bounded subset of events in a given stream for performing analysis. A window can be sliding or batch. Each such window can be time or length (count) based. A time-based window can be used to perform a computation on events that arrived in a specific interval of time such as the last T seconds. A count-based window is used to perform a computation based on the values of the last C events where C is the specified count (length) parameter.

Irrespective of whether it is time-based or length-based a window has an associated expiration policy: sliding or batch. A sliding window is characterized by a window size and a window slide parameter. If the window size and the slide parameter for a time-based sliding window, for example, are T and S ($T \leq S$), then the window captures events that occurred in the last T units of time. Every S unit of time, the previously occurred events inside the window are flushed. For a batch policy, the adjacent windows are nonoverlapping and the processing is done when the batch is complete. For example, a time-based batch window of the duration of T units of time will perform computations every T units of time on the events received within the window. The events after processing will be flushed.

Query Execution Plan: The query execution plan is the representation of a query within the CEP engine. It depends on the type of the query. There are two query types: regular and pattern. For regular queries, the order of arrival of events is not important and the result of execution is independent of the order in which events arrived. For pattern queries, the order of event arrival is important and this order influences the results of the query execution.

Event time stamping: Recording the time of events is performed by a CEP engine. The times associated with an event are described next.

- *RE Generation Time*: is the local time at which the event is generated and is typically recorded by the event generator itself.
- *RE Ingestion Time*: After generation, the RE is processed by the CEP engine and the time at which the RE is ingested by the CEP engine is called the RE ingestion time.
- *CE Detection Time*: is the time at which the CE is detected.

8.6.3 Example CEP Systems

CEP has a strong similarity with active databases (Rizvi 2005). An active database can respond automatically to a continuous stream of events using Event Condition Action rules. Such rules perform the desired operations when a certain logical condition is met as REs arrive on the active database system. A CEP system, however, has a number of advantages over active databases. It addresses several limitations of active databases, including lack of standardization and difficulty in performing various distributed and parallel operations.

Early works on CEP are available in (Luckham 2001; Motwani et al. 2002). CEP is often defined by authors as a set of rules and techniques to analyze and control event-driven information processing systems that can range from large internet-based to smaller cellphone-based systems.

The initial work on CEP includes the Rapide event-driven simulation language that can be used to model multilayered event-based systems (Luckham 1998). Another early work on CEP is available from (Motwani et al. 2002) that concerns the Stream project, a system for processing continuous queries on multiple continuous data streams. The system supports a declarative query language and can handle high data rates and query workloads providing approximate answers when the available resources are limited.

A CQL called SASE is proposed in (Gyllstrom et al. 2006) for processing radio frequency identification (RFID) data. SASE uses a tree-based approach to perform CEP. Another popular CEP engine is Esper (Microsoft Corporation n.d.). It also provides the core functionality in Oracle CEP. One of the limitations of Esper is that it is a centralized CEP system. However, it has been adapted to integrate it with the distributed/parallel processing machinery of Apache Storm (Al-Haboobi et al. 2016). Another addition to open-source CEP platforms is Apache Flink (Carbone et al. 2015), which can be deployed in a cluster mode. An attractive feature of Flink is its ability to handle out-of-order events by using time stamps and watermarks.

8.6.4 CEP Platforms

A number of platforms for performing CEP is available. One such platform that is used in the system described in Section 8.7 is discussed next.

8.6.4.1 Apache Siddhi

Apache Siddhi (Suhothayan et al. 2011) is a lightweight CEP engine with a relatively small memory footprint in comparison to other stream processing systems. It is used for stream processing applications that require CEP. It provides support for a CQL for query processing. It has also been ported to smaller devices, such as Android-based smartphones and Raspberry Pi devices (Upton and Halfacree 2016). It can receive streaming data including sensor data streams. The input data can come from various event sources and the detected CEs can be sent to various event sinks. Examples of such event sources and sinks include MQTT, Kafka, TCP, and HTTP. The sensor data formats and the formats of data in the CEP alerts generated after CEP include Binary, JSON, Text, and XML.

8.7 Remote Patient Monitoring System

The description of a system that performs CEP is discussed in this section. The RPM system described concerns a chronic cardiac patient fitted with body sensors that monitor various statistics associated with the patient, such as their heart rate and respiration rate (Dhillon et al. 2021). The outputs of the sensors generate REs streams that are analyzed by CEP engine. The engine tries to find patterns that predict impending health problems for this patient. Two versions of the RPM system are discussed: the Server CEP (SCEP) system and the Mobile CEP (MCEP) system (see Section 8.7.2). A schematic diagram of the three-tiered SCEP system is presented in Figure 8.14. The sensors communicate with the

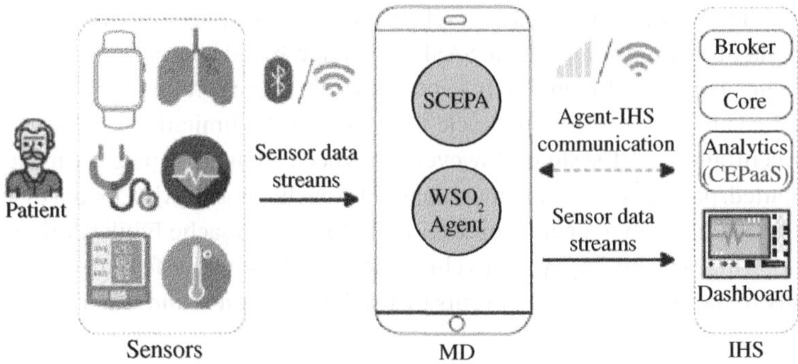

Figure 8.14 The architecture of the system performing CEP at the backend. Source: Dhillon et al. (2018)/with permission of IEEE.

gateway agent through Bluetooth or Wi-Fi, which is a mobile device (MD) such as a smartphone. This gateway agent runs a SCEP application SCEPA and a WSO_2 (Dahanayakage et al. 2018) application communicates with a server called the IoT Hospital Server (IHS) (Dhillon et al. 2021) that detects CEs from the sensor streams received from MD. Communication between IHS and MD is achieved typically through a cellular connection. IHS runs the Analytics CEP as a service (CEPaaS) service and displays the results on a dashboard. A similar architecture can be used in a smart infrastructure management system in which the wireless sensors can be replaced by wired sensors in a smart building or smart home, for example, that can sense the light intensity and temperature in specific rooms that are the subject of analysis. MD in such a situation can be replaced by a local server. SCEP uses a socket to receive sensor data streams and an MQTT service to send the data to IHS. Since multiple sockets are in use for handling multiple sensors, a thread-safe queue is placed between the sockets and the MQTT service that forwards the data to the IoT server.

8.7.1 The SCEP System

The SCEPA system is described with the help of Figure 8.15. Note that a solid line represents multiple data streams whereas a dashed line represents a single data stream.

Each component of SCEPA is briefly described.

(1) MQTT Receiver: Each receiver receives a raw sensor stream. Multiple instances of MQTT receivers are used to receive multiple streams, one for each sensor.
(2) Arrival Time Stamping: Each time stamping component receives a particular stream and appends a system-generated time stamp to capture the stream arrival time.

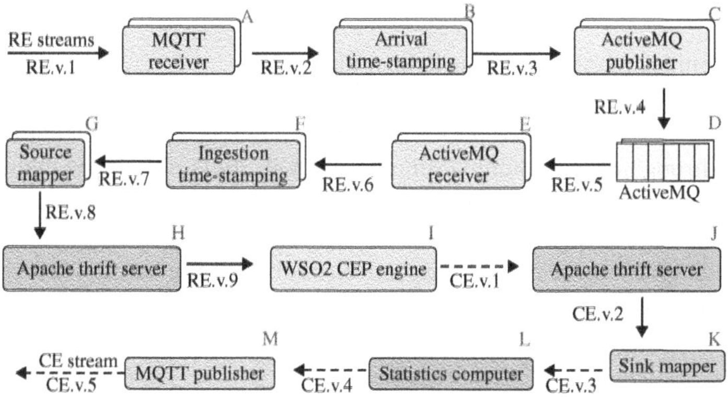

Figure 8.15 The SCEPA system. Source: Dhillon et al. (2021)/Springer Nature.

(3) Active MQ Publishers (Dhillon et al. 2021): are used as queues associated with Java Message Service (JMS). Each active MQ publisher sends the message to a specific brokered queue managed by an Active MQ broker.
(4) Active MQ Receiver: receives sensor data streams from a specific Active MQ. The Active MQ protocol provides support for the Advanced Message Queuing Protocol (AMQP), which is a protocol for message-oriented middleware.
(5) Ingestion Time Stamping: This is performed by using the system clock to append the ingestion time before sending the data to the CEP engine. Multiple ingestion time components work in parallel for time stamping the different sensor streams.
(6) Source Mapper: The responsibility of this component is to map the format of the sensor data stream events to the format that can be handled by the CEP engine.
(7) Apache Thrift Server: furnishes a remote procedure call (RPC) framework to provide interoperability among different languages used to build cross-platform services.
(8) CEP Engine: generates a single CEP stream from the multiple sensor data streams and sends it to the sink mapper. In addition, the CEP engine generates the CE detection time stamp.
(9) Apache Thrift Server: receives the CEs and sends them to the sink mapper. Sink Mapper: performs the format conversion for the data so that it can be processed by the Data Analytics Server.
(10) Statistics Computer: Various metrics used in performance analysis are computed by this component by using the various time stamps generated in the various stages of the workflow.
(11) MQTT Publisher: publishes the various streams to the event listeners such as a dashboard and a database.

8.7.2 The Mobile CEP (MCEP) System

The MCEP system is described with the help of Figure 8.16.

As in the case of the SCEP system, the MCEP system also has a three-tiered architecture. The CEs are detected on the MD using an embedded CEP engine (implemented through Apache Siddhi (Suhothayan et al. 2011). This is the main difference between SCEP and MCEP. In case of MCEP, the CE stream is generated on MD running the MCEP Application, MCEPA, and a WSO2 Agent is sent to IHS. The MCEP system has an event listening service running on the hospital server IHS which subscribes for the CE stream generated on MD.

8.7.3 Performance Analysis

The performance of the MCEP system is compared with that of the SCEP system. Two performance metrics are used: Remaining Battery Life (RBL) and CEP latency, L. Each of them is discussed separately. The details of the experiments conducted for doing the performance analysis are not required for the current discussion and are available in (Dhillon et al. 2021).

8.7.3.1 Remaining Battery Life

An experiment monitoring the power consumption of the systems was performed for a period of 60 minutes. A Google Pixel mobile smartphone running the Android Nougat operating system was used as the MD. The RBL for both the SCEP and MCEP systems is presented in Figure 8.17 for different values of T_{run} that correspond to the time elapsed after the start of the experiment. There are two sets of results for both SCEP and MCEP. SCEP–FG corresponds to the SCEP system when the SCEP application was running in the foreground (with no applications running in the background) while SCEP–BG corresponds to the SCEP system when the SCEP application was running in the background (with

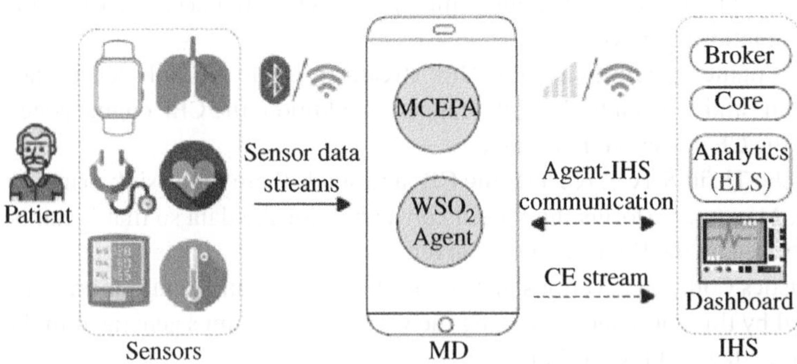

Figure 8.16 The MCEP system. Source: Dhillon et al. (2021)/Springer Nature..

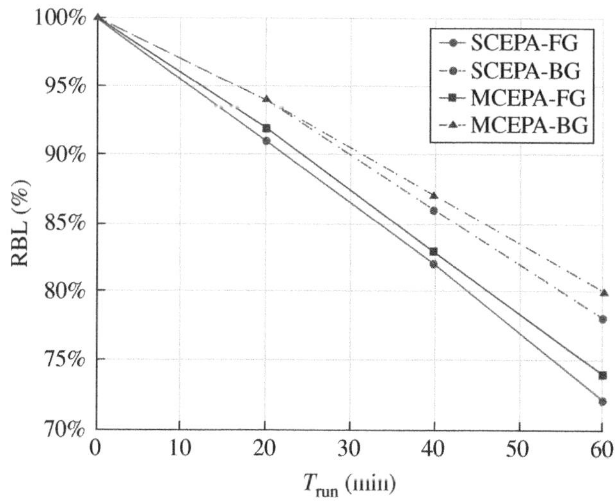

Figure 8.17 Comparison of battery usage by the different systems. Source: Dhillon et al. (2021)/Springer Nature.

no applications running in the foreground). Similarly, MCEP–FG corresponds to the MCEP system when the MCEP application was running in the foreground (with no applications running in the background), while MCEP–BG corresponds to the MCEP system when the MCEP application was running in the background (with no applications running in the foreground).

For a given scenario (FG or BG), the battery usage for the MCEP application is lower than that of the SCEP system. Also, for a given system (SCEP/MCEP) the battery usage for FG is lower than that observed for BG (see Figure 8.17).

8.7.3.2 CEP Latency

The latency associated with the detection of a CE is an important concern in real-time systems especially for the RPM system since this latency is an important component of the time taken by the health services team to respond to an impending health problem. Experiments were conducted by using a synthetic workload deployed on a Google Pixel MD running Android Nougat. A detailed description of the experiments and the parameters are provided in Dhillon et al. (2021).

A CE is generated when CQL is able to find a given pattern match from multiple sensor data streams. The latency associated with a CE is measured as the difference between the time the CE is detected and the time of ingestion for the first event from any sensor stream that leads to the CE. A simulated sensor stream was used in the latency measurement experiment. The mean latency over all CEs detected

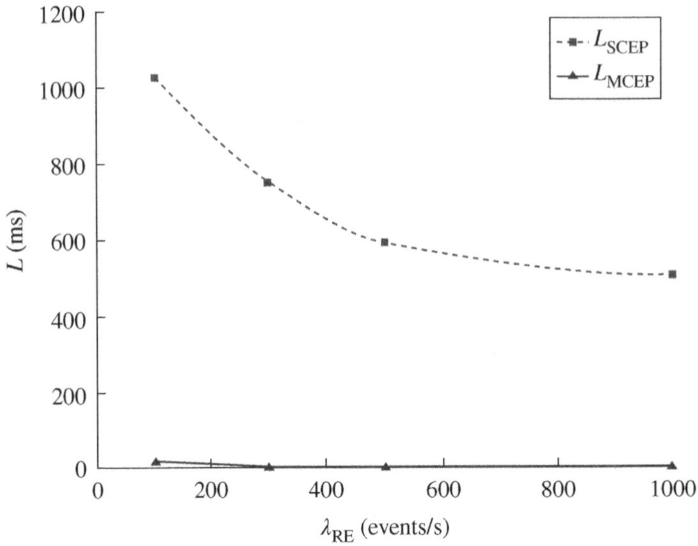

Figure 8.18 Comparison of latency between SCEP and MCEP. Source: Dhillon et al. (2021)/Springer Nature.

during the experiment referred to as L_{MCEP} for the MECP system and as L_{SCEP} for the SCEP system are shown in Figure 8.18. λ_{RE} is the rate at which events from the simulated sensor streams arrive on the system. For any given λ_{RE}, the latency produced on the MCEP system (L_{MCEP}) is orders of magnitude smaller than the latency produced on the SCEP system (L_{SCEP}).

Overall, the MCEP system is observed to be superior to the SCEP system, which has a number of limitations that include the following.

- It necessitates the MD to remain connected to the hospital server at all times.
- It gives rise to an increase in user cost as all the raw sensor data streams have to be forwarded to the remote server.
- If SCEP is handling multiple clients, there can be queuing delays on the server side.
- Since data packets can be transmitted in multiple sessions, SCEP can lead to out-of-order delivery of various sensor streams with respect to one another.

8.8 Summary

This chapter focused on streaming systems in which streams of data tuples are processed. Two types of stream processing techniques are discussed: priority scheduling and CEP. The workload on each of these systems is dependent on the contents of the data that arrive in the system. The dynamic data-driven scheduler discussed earlier in this chapter adjusts the priority of data elements based on the existence of HP keywords in the data stream. The algorithmic components of the scheduling technique are discussed. The DDPS technique can be used in conjunction with the Spark Streaming system. Priority-based resource management techniques for multitenant storm clusters are also discussed. The resource manager on such systems needs to distribute the resources among multiple storm topologies based on their respective priorities. For the static priority scheduler, the priorities of the topologies are fixed and are specified before run time. In the case of DPSs, a topology priority can change dynamically at run time. The priority of a topology is changed when the topology sends a "trigger ON" signal to the PM. Both an algorithmic description and performance analysis of each of these static and dynamic priority-based resource management techniques are discussed.

CEP discussed in this chapter forms the second focus of this chapter. The efficacy of the technique is demonstrated through a performance analysis of a prototype deployed on Apache Siddhi. A CE results from the combination of multiple REs arriving in the system. Various terminologies used in the context of CEP are introduced and important components of the CEP architecture are discussed. An RPM system that is based on a CEP engine provided by the Apache Siddhi system is discussed. Two versions of the system are described. In the SCEP system, the CEP is performed at the backend while CEP is performed at the edge on a MD for the MCEP system. A discussion of the performance comparison of the two systems is included.

Exercises

1 Extend the DDPS algorithm described in this chapter to handle more than two priority levels.

2 Explain what is meant by a multitenant system. Discuss the challenges it poses to resource managers. Explain how the isolation scheduler tries to address these. Include the limitations of isolation scheduler in your discussion.

3 Compare the DPS and SPS techniques described in the context of the storm. Include the overheads associated with each technique.

4 Discuss the relative merits and demerits of the SCEP and MCEP systems. Include a discussion of the performance of the two systems as captured in Figures 8.17 and 8.18.

5 Discuss how the MCEP system can be extended to a hybrid system in which real-time analytics is performed on the MD and long-term predictive analytics that uses data captured over a period of time at the backend server.

6 One problem with the MCEP system is the MD running out of power. Discuss how the system described in this chapter can be enhanced to handle such a situation.

References

D.J. Abadi, D. Carney, U. Cetintemel, M. Cherniack, C. Convey, S. Lee, M. Stonebraker, N. Tatbul, S. Zdonik, "Aurora: A New Model and Architecture for Data Stream Management," *The International Journal on Very Large Data Bases (VLDB) Journal*, Vol. 12, Issue 2 2003, pp. 120–139.

T. Ajila, S. Majumdar, "Data Driven Priority Scheduling on a Spark Streaming System", *IEEE/ACM Intl. Conference on Cluster, Cloud and Grid Computing (4th IEEE/ACM International Workshop on Distributed Big Data Management)* Larnaka (Cyprus), May 2019.

A.S. Al-Haboobi, A.F. Alharan, R.H. Alsagheer, "Experimenting with Storm/Esper Integration and Programmatic Generation of Storm Topologies," *International Journal of Computer Science and Information Security*, Vol. 14, Issue 8, 2016, pp. 169–178.

Apache Software Foundation, "Kafka: A Distributed Streaming Platform", n.d., https://kafka.apache.org/. [accessed: 23-April-2018].

"Apache Storm", n.d.-a, https://spark.apache.org/. [accessed 2021].

Apache Spark, "RDD Programming Guide", n.d.-b, http://spark.apache.org/docs/3.0.1/rdd-programming-guide.html [accessed: 12-Jan-2021].

Apache Spark, "Structured Streaming Programming Guide", n.d.-c, http://spark.apache.org/docs/3.0.1/structured-streaming-programming-guide.html. [accessed: 02-Mar-2021].

Apache Spark, "Job Scheduling", n.d.-d, https://spark.apache.org/docs/latest/job-scheduling.html [accessed: 20-Aug-2022]

Apache Storm, "Scheduler", n.d.-a, https://storm.apache.org/releases/1.2.4/Storm-Scheduler.html. [accessed: 2023].

Apache Storm, n.d.-b, https://storm.apache.org.

R. Birke, M. Bjorkvist, E. Kalyvianaki, L.Y. Chen, "Meeting Latency Target in Transient Burst: A Case on Spark Streaming," *Proceedings of the IEEE International Conference on Cloud Engineering (IC2E)*, Vancouver, Canada, 2017, pp. 149–158.

P. Carbone, A. Katsifodimos, S. Ewen, V. Markl, S. Haridi, K. Tzoumas, "Apache Flink: Stream and Batch Processing in a Single Engine," *Bulletin of the IEEE*

Computer Society Technical Committee on Data Engineering, Vol. 36, Issue 4, 2015, pp. 28–38.

Chakraborty, R., Majumdar, S., "Priority Based Resource Scheduling Techniques for a Resource Constrained Stream Processing System", *Proc. the 4th IEEE/ACM International Conference on Big Data Computing, Applications and Technologies (BDCAT)*, Austin, United States, December 2017.

D. Cheng, Y. Chen, X. Zhou, D. Gmach, D. Milojicic, "Adaptive Scheduling of Parallel Jobs in Spark Streaming," *Proceedings of the IEEE Conference on Computer Communications (INFOCOM)*, Atlanta, USA, 2017, pp. 1–9.

T. Dahanayakage, S. Suhothayan, M. Dayarathna, "Distributed Stream Processing with WSO2 Stream Processor", White Paper, 2018, https://wso2.com/whitepapers/distributed-stream-processing-with-wso2-stream-processor/

T. Das, Y. Zhong, I. Stoica, S. Shenker, "Adaptive Stream Processing using Dynamic Batch Sizing," *Proceedings of the ACM Symposium on Cloud Computing (SOCC '14)*, Seattle, USA, 2014, pp. 1–13.

DeZyre, "Top 5 Apache Spark Use Cases", n.d., https://www.dezyre.com/article/top-5-apache-spark-use-cases/271. [accessed: 01-Mar-2021].

A. Dhillon, "An Edge Computing-based Complex Event Processing Technique for Sensor-based Systems", M.A.Sc. Thesis, Department of Systems and Computer Engineering, Carleton University, Ottawa, Canada, 2018.

A. Dhillon, S. Majumdar, M. St HIlaire, A. El-Haraki, "MCEP: A Mobile Device Based Complex Event Processing System for Remote Healthcare", *The 11th IEEE International Conference on Internet of Things (IThings)*, Halifax, Canada, August 2018.

A. Dhillon, S. Majumdar, M. St-Hilaire, A. El-Haraki, "Complex Event Processing in Sensor-Based Environments: Edge Computing Frameworks and Techniques", in *Mobile Edge Computing* (eds: A. Mukherjee, D. De, S.K. Ghosh, R. Buyya), Springer, 2021.

D. Gyllstrom, E. Wu, H.J. Chae, Y. Diao, P. Stahlberg, G. Anderson, "SASE: Complex Event Processing over Streams," *CoRR*, Vol. abs/cs/0612128, 2006.

U. Hunkeler, H.L. Truong, A. Stanford-Clark, "MQTT-S — A Publish/Subscribe Protocol for Wireless Sensor Networks," *Proc. of the 3rd International Conference on Communication Systems Software and Middleware and Workshops (COMSWARE)*, 2008, pp. 791–798.

D. Lau, J. Liu, S. Majumdar, B. Nandy, M. St-Hilaire, C.S. Yang, "ACloud-Based Approach for Smart Facilities Management", *Proc. 2013 IEEE Conference on Prognostics and Health Management (PHM)*, Gaithersburg, USA, June 2013.

D.C. Luckham, "Rapide: A Language and Toolset for Causal Event Modelling of Distributed System Architectures," *Proc. of the Worldwide Computing and Its Applications (WWCA)*, 1998, pp. 88–96.

D.C. Luckham, *The Power of Events: An Introduction to Complex Event Processing in Distributed Enterprise Systems*, Addison-Wesley Longman Publishing Company, 2001.

S. Majumdar, "Leveraging Cloud Computing and Sensor-Based Devices in the Operation and Management of Smart Systems", *Smart City: Software Services and Cyber Infrastructure* (eds: M. Maheswaran, E. Badidi), Springer, 2018.

Microsoft Corporation, "Microsoft StreamInsight", n.d., https://docs.microsoft.com/en-us/previous-versions/sql/streaminsight/ee362541(v=sql.105). [accessed: 8-May-2018].

R. Motwani, J. Widom, A. Arasu, B. Babcock, S. Babu, M. Datar, G. Manku, C. Olston, J. Rosenstein, R. Varma, "Query Processing, Resource Management, and Approximation in a Data Stream Management System," *Proc. of the Biennial Conference on Innovative Data Systems Research (CIDR) Stanford InfoLab*, 2002, pp. 1–12.

S. Rizvi, "Complex Event Processing Beyond Active Databases: Streams and Uncertainties", 2005, https://www2.eecs.berkeley.edu/Pubs/TechRpts/2005/EECS-2005-26.pdf. [accessed: 8-August-2018].

S. Suhothayan, K. Gajasinghe, I. Loku Narangoda, S. Chaturanga, S. Perera, V. Nanayakkara, "Siddhi: A Second Look at Complex Event Processing Architectures," *Proc. of the ACM Workshop on Gateway Computing Environments (GCE)*, 2011, pp. 43–50.

E. Upton, G. Halfacree, *Raspberry Pi® User Guide* (4 ed.), John Wiley & Sons Publishers, 2016.

S.B. Zdonik, M. Stonebraker, M. Cherniack, U.C. Cetintemel, M. Balazinska, H. Balakrishnan, "The Aurora and Medusa Projects," *IEEE Database Engineering Bulletin*, Vol. 26, Issue 1, 2003, pp. 3–10.

9

Data Indexing and Filtering Techniques for Big Data Systems

9.1 Introduction

Data are sometimes referred to as the "oil" of today. Some people speculate that data are soon going to replace oil as the world's most valuable resource. A large volume of raw data are produced every day. An example is Facebook, which is said to be processing approximately 500 Bytes of data (Kambatla et al. 2014). A Boeing jet engine is capable of producing several TBytes of data every 30 minutes of its flying time. According to Sayce (n.d.), as of May 2020, 500 million tweets are reported to be generated every day. There are many more examples of large data that is produced regularly. This includes scientific journals, newspapers, and medical records. However, like crude oil this raw data often referred to as big data needs to be processed to extract meaningful information from the underlying data set. A number of characteristics of big data pose challenges to this information extraction process. They include the following (Analytics Insight 2021).

Volume: refers to the large size of the data that can range from Mbytes to Gbytes to Pbytes and more. Analyzing such data requires a powerful underlying processing engine.

Velocity: refers to the speed at which data are generated and the time at which the results of analysis are required to be produced. The expected duration of performing the data analytics can vary from an hour to minutes to seconds and even milliseconds. Thus, depending on the use case appropriate real-time and near real-time techniques need to be employed to meet the desired objectives.

Variety: refers to the type of data that comes in various forms including text, numeric, video and audio. Each of these categories may be divided into subcategories that include the format of the data (e.g. XML and JSON), the type of coding (e.g. MPEG) applied to the data file, and so on. The existence of various data types makes data processing more complex.

Veracity: refers to the accuracy and truthfulness of data. This attribute of a given body of data may depend on the data source and the process used to produce it.

Resource Management on Distributed Systems: Principles and Techniques,
First Edition. Shikharesh Majumdar.

If the data are produced by multiple sources, then the problem of determining the veracity of the data are harder since it will now be determined by the data accuracy produced by each of these sources.

Effectively utilizing the potential of big data for increasing revenue is discussed in Section 9.2, which identifies two techniques, reducing search latency and reducing data volume, that are introduced in Sections 9.3 and 9.8, respectively. The reduction of search latency through data indexing is discussed first. Section 9.4 focuses on inverted index-based techniques, whereas Section 9.5 concerns graph-based indexing (GBI) techniques. Handling Boolean AND queries for searching data are discussed in Section 9.6, whereas performance analyses of the indexing techniques are covered in Section 9.7. Both the relationship between search time and the various system and workload characteristics, as well as the system overheads associated with indexing are discussed.

Section 9.8 focuses on data filtering techniques used for reducing data volume that, in turn, reduces data search time. This is achieved by techniques for filtering data based on users' preferences. Data filtering itself is a computation intensive operation and thus uses parallel processing frameworks to reduce the filtering time. Parallel processing platforms that run the filtering algorithms are discussed in Section 9.9. Use cases for the data filtering algorithms are discussed in Section 9.10. Data filtering can be performed both on stored data as well as on streaming data. Techniques used for filtering stored data are described in Section 9.11, whereas filtering of streaming data are discussed in Section 9.13. Section 9.12 focuses on the performance analysis of stored data filtering techniques, whereas Section 9.13 includes a discussion of streaming data filtering techniques and their performance. User preferences characterized by Boolean operators such as AND and OR are discussed in Section 9.14, which includes a description of system performance for such use cases. Section 9.15 presents a short summary of the chapter highlights.

9.2 Harnessing Big Data

Data are recognized as an important asset for a company. Data mining is used to make important management decisions such as planning for future products, analyzing employee productivity, and so on. Data can thus be considered as a source of revenue for enterprises. Like many other computing infrastructures data are also being provided to clients as a service. Like all other computing resources, data needs to be effectively managed. One method of efficient processing of large volumes of data are to speed up the data processing operations. Using parallel processing frameworks such as Hadoop/MapReduce and Spark are discussed in this book. Another way to speed up processing is to improve the latency of data searches performed on large data repositories. Two different classes of techniques, *decreasing search latency and reducing data volume,* are discussed in this book. Speeding up processing is often achieved by using parallel processing frameworks such as MapReduce (discussed in Chapter 6) and Spark (discussed in this chapter). Decreasing search latency and reducing the data volume are discussed in this chapter. Data indexing for achieving faster searches is discussed first in Section 9.3. This is followed by a discussion of selective data filtering techniques that can reduce the volume of data to be processed (see Section 9.8).

9.3 Data Indexing

Indexing is a technique for reducing the latency of performing searches on a document set with a large volume. By using indexing in the context of a product catalog, for example, all the different items in the same class (such as "laptop" or "skillet") can be retrieved quickly. When the user uses a search query involving "skillet" in the products catalog of a vendor all the items in that category are displayed. Search queries can be exclusive keyword queries that involve only single keywords. No other keyword can be present in the items returned as a result of a search. In addition to exclusive keyword queries, Boolean queries are often used. A Boolean query comprises multiple keywords that are connected by one or more Boolean operators such as OR, AND NOT. Of particular interest is the Boolean AND query that is often used in search engines such as Google. Single-word queries are often not very useful because they may return a large set of documents with many items in the list being of no interest to the user. A Boolean query, on the other hand, can be used to narrow the search.

An index is used to find the list of documents that contain a specific keyword. Note that all the words in a document are not indexed. Only a certain set of words identified in a keyword extraction step (Mahapatra and Biswas 2011) are associated with an index. This step is performed prior to the keyword indexing step. Two types of indexing are discussed. The inverted index is a well-known technique used in many search engines, such as Google. This technique is described in the following section, whereas GBI is discussed in Section 9.5.

9.4 Inverted Index

Inverted index is a data structure used in information retrieval. It maps a keyword to a list of documents, often referred to as a *postings list*. Examples of documents are web pages, tweets, or other textual documents. Each document can be identified by a unique ID. The postings list is a list of all the documents with IDs that contain the specific keyword. Consider the following example. The document set comprises eight documents with IDs D1–D8. There are three keywords to be indexed "Anthony", "Caeser," and "Brutus".

Anthony → D1,D2,D3,D4

Brutus → D1,D2,D5,D6

Caesar → D1,D6,D7,D8

Figure 9.1 Mapping between index and postings list for the example.

The keyword "Anthony" appears in documents D1–D4; the keyword "Brutus" appears in documents D1, D2, D5, D6; and the keyword "Caeser" appears in documents D1, D6–D8. The inverted index captures the mapping between the keywords and the postings list (see Figure 9.1).

There are different ways to implement an inverted index (Mahapatra and Biswas 2011). The first includes a list of keywords and the corresponding postings list (see Figure 9.1). The second also includes the frequency of occurrence of each keyword in each document along with the respective ID. In the third, the positions of each keyword in the document, along with its document ID, are recorded.

Hash tables are often used in the context of an inverted index. Recall that a hash table is based on the notion of a set of key-value pairs. A key is a string and is converted into an integer using a hash function. This integer can then be used to index into an array and store the corresponding value. For an inverted index, each keyword is a string and the corresponding value is its postings list.

One way to handle Boolean AND queries is to look for each keyword in the query and, obtain the postings list for each and then perform the desired intersection logical operations among the posting lists. The documents identified by the document IDs are obtained as a result of the intersection operation. With large volumes of textual documents this can significantly increase the result computation time. A graph-based technique that can detect the simultaneous occurrence of multiple keywords in a document and reduce this overhead significantly is described in Section 9.5.

9.4.1 Enhancements to the Inverted Index Technique

One of the problems of Inverted Index is the large time for the execution of a search query when the keyword does not exist in any postings list. A technique called bloom filter has been combined with an inverted index (Kumar et al. 2014) to enhance the performance of search queries for such use cases.

Another variant of the original Inverted Index technique is the Generalized Inverted Index (Ginix) (Wu et al. 2013). The difference between Ginix and the original Inverted Index lies in the postings list. Consecutive document IDs are represented as intervals in Ginix. For example, if there are four documents with

IDs 1–4 containing the keyword Computer, the postings list for Computer will contain the entry [1,4]. The union and intersection operations on the postings lists are performed by directly using the intervals and not expanding the list. This tends to reduce the memory consumption for storing the postings list.

Additional parameters are sometimes included in Boolean AND, OR, and NOT queries. In addition to the search keywords, a proximity parameter is specified (Mikolov et al. 2013). This parameter specifies how close the keywords should appear in the document to qualify for being in the resultant set. The closeness may be specified in terms of words, sentences, or paragraphs.

A bit map-based technique for optimizing the execution of Boolean AND queries is presented in Fontoura et al. (2011). The technique uses a bit map to indicate the co-occurrence of keywords in documents. Each bit map is an array of 1s and 0s. A bit map containing X bits is attached to each document ID in the document list of keyword Y. The bit map indicates which other frequently occurring keywords in the set of X keywords co-appear with keyword Y. The advantage of the technique is that only the postings list of Y will be enough to handle an AND query involving keyword Y and any other X keywords. The result of the AND query will contain all the document IDs in the postings list of Y for which the bit vector contains a 1 in the position of the other X search keyword(s).

Elastic Search is a well-known search engine built using Apache Lucerne (Apache Software Foundation n.d.). It is used by many enterprises, such as Netflix and Tinder, for performing text searches. It is built using the inverted index as the indexing structure.

9.5 Graph-Based Indexing

The GBI technique uses a directed graph to index the keywords appearing in the documents. Such a graph comprises a set of nodes connected by directed edges. The label of each node is one of the keywords to be indexed. If the keyword i and keyword j appear in the same document an edge is placed between the two nodes. The direction of the edge is based on the lexicographical ordering of node labels. The edge points from node i to node j because i appears before j in the English alphabet. The label of an edge comprises the document IDs of both the documents that correspond to the nodes at the head and tail of the edge.

Consider, for example, two documents with ID 1 and ID 2. The keywords appearing in each of the documents are presented in Table 9.1.

The graph structure corresponding to the GBI technique is presented in Figure 9.2.

There are three nodes in the graph, one for each keyword: Clinton, Trump, and Justin. The edge with labels 1 and 2 signifies that both Clinton and Trump appear

Table 9.1 Example of data comprising two documents.

Document ID	Keywords
1	Justin, Trump, Clinton
2	Clinton, Trump

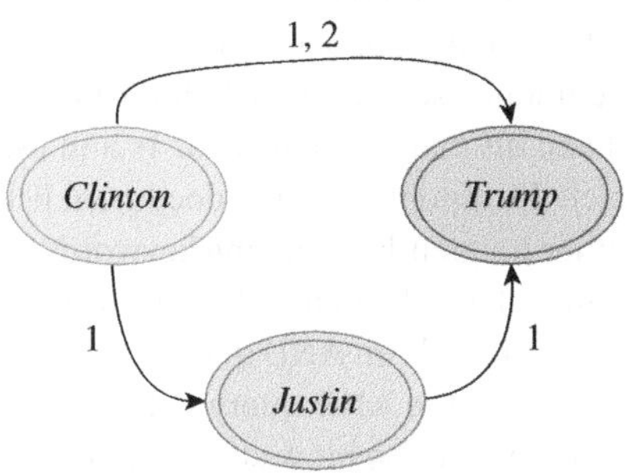

Figure 9.2 GBI for the two document example. Source: Mohideen et al. (2020)/with permission of IEEE.

in documents with IDs 1 and 2. The edge is directed from the node with the label Clinton to the node with the label Trump because of the lexicographical ordering used. The label between Clinton and Justin is labeled 1 because both of the keywords appear in the document with ID 1. The edge between Justin and Trump is also labeled 1 because Justin and Trump appear in the document with ID 1.

A modified version of the well-known adjacency matrix is used to represent a GBI. Only the keywords that have occurred together in a document are stored. Note the difference with a conventional adjacency matrix to represent a graph in which even the nodes that do not have an edge between them are stored. By not storing all the nodes in the graph, the modified adjacency matrix leads to a smaller space overhead. A two-dimensional hash table is used to store such a modified adjacency matrix. Two keywords that appear together in a document are grouped together to serve as a key for the hash table, whereas the document IDs of these documents that contain these keywords serve as the value corresponding to the key.

A high-level algorithm for constructing a GBI when there are at least two keywords in each document is presented in Figure 9.3. The GBI for a given document set is stored in a hash table G. As and when a new document is to be processed this algorithm is called after the keyword extraction process is complete. A set of keywords (S) and the ID of the document (Id) are the inputs for the algorithm and the new or updated hash table for the GBI structure produced by the algorithm is stored in G. S is sorted in ascending order (line 1). Set S is processed iteratively

Figure 9.3 High-level algorithm for the construction of GBI. Source: Mohideen et al. (2020)/with permission of IEEE.

```
Input: Set (S) of keywords and identifier (Id) of
       document
Output: New or updated hash table representing the
        Graph-Based Index (G)
Condition: At least two keywords exists in S
1  Sort set S in ascending order
2  foreach keyword key1 in list S do
3  |  if key1 is last entry in list S then
   |  |  /* Indexing process is complete */
4  |  |  break
5  |  end
   |  /* When a document has more than one keyword
   |     to index */
6  |  foreach keyword key2 occurring after key1 do
7  |  |  Key| ← hash table key [key1][key2]
8  |  |  If an entry identified by Key exists in G,
   |  |    append I d to the list of values, otherwise
   |  |    create an entry using Key with I d as value.
9  |  end
10 end
```

(lines 2–9), with each keyword being handled in a separate iteration. If the current keyword is processed, and key 1 is the last keyword in the set, it means that there is no keyword following this keyword. Since G should have entries for only pairs of keywords appearing together in a document, this indicates that the GBI construction is complete, and the algorithm terminates (line 4). Each keyword, key 2, appearing after key 1 in S, is then processed separately in lines 6–9. If the hash table entry [key 1, key 2] already exists in G, the document identifier (Id) is appended to the list of values identified by this hash table entry. Otherwise, a new entry is created for the key 1, key 2 pair as a new key for the hashable and ID is added as the value field. The procedure continues until all the entries in S are processed.

Documents with a single keyword were not handled in the algorithm given in Figure 9.3. In such a case, a looping edge that starts with and ends on the same keyword node is added. Consider, for example, four documents. The keywords and document IDs are presented in Table 9.2.

The GBI for this case is presented in Figure 9.4.

Note that since document 4 contains only the keyword C, the looping edge with label 4 is added to the node labeled C.

9.6 Boolean AND Queries

Boolean AND queries are often used in the context of various applications such as Google searches, social media, and database systems. In searches performed on a search engine such as Google AND queries are often used to narrow down

Table 9.2 Example with four documents.

Document Id	Keywords
1	A, B, C
2	B, C
3	A, C
4	C

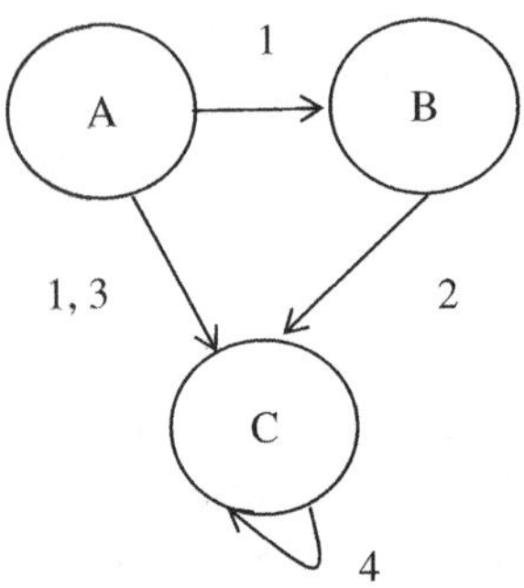

Figure 9.4 GBI for document set captured in Table 9.2.

the search results to documents identified by relevant links (Zhang et al. 2008). Using only single keywords in searches often generates a very large number of documents that may take a long time for the user to search through and using multiple keywords in the search can significantly reduce the number of documents resulting from the search. Using a GBI technique to perform such AND searches is discussed in this section.

The algorithm for performing an AND search that is used to identify the documents that contain all the keywords provided as a set of arguments to the search method is presented in Figure 9.5 (indicated as Algorithm 2 in the figure from Mohideen et al. 2020). S, the set of keywords of interest (the number of which is more than or equal to two), is provided as the first input to the algorithm. The second input parameter is the GBI hash table G.

Upon start, the algorithm sorts the set S in an ascending lexicographic order (line 1), and the number of search keywords in S is determined (line 3). Each keyword key 1 in S is processed iteratively by using a loop that spans lines 4–21. The postings list is collected only if key 1 is not the last keyword, and it does not appear with another keyword pair that has already been processed during the iteration. The rationale for this is explained briefly with the help of an example. Consider a set of search keywords: A, B, C, and D. There is a posting list for each of the keyword pairs that appear together in one or more documents. Each entry

Figure 9.5 Algorithm for performing an AND search. Source: Mohideen et al. (2020)/with permission of IEEE.

Algorithm 2: Boolean AND Search using GBI

```
Input: Set of search keywords (S), hash table
       representing the Graph-Based Index (G)
Output: NULL or the matching document ids.
Condition: At least two unique keywords exists in S
Sort S in ascending order
postings_list_collection ← an empty hash table
search _count← size of list S
foreach key word key1 in S do
    if key1 is not the last entry in S then
        /* Check if all the search keywords are
           connected to each other in GBI */
        foreach keyword key2 occurring next to key1
         in S do
            Key ← hash table key [key1][key2]
            if Key exists in hash table G then
                if Key is non-overlapping or last
                 keyword pair in S then
                    /* Collect selected postings
                       list */
                    create entry in
                     postings_list_collection with
                     key1 as key and postings list at
                     G[Key] as value
                end
            else
                postings_list_collection ← NULL
                /* If any of the search keywords
                   are not connected, than
                   terminate keyword relationship
                   checking. No possibility of
                   common documents */
                break
            end
        end
        if postings_list_collection is NULL then
            /* Terminate search process */
            return NULL
        end
    end
end
if postings_list_collection is not NULL then
    /* If all the search keywords are connected
       to each other than common documents are
       possible */
    if search_count equals 2 then
        /* If only two search keywords then
           return result directly */
        return postings_list_collection
    else
        Perform Small-Vs-Small (SvS) intersection
         algorithm on postings_list_collection as
         described in Algorithm 3
    end
end
```

in *G* corresponds to one such pair. If two keywords do not appear together in any document, then there is no entry for this keyword pair in *G*. Now consider the case in which *G* includes entries for AB, AC, AD, BC, BD, and CD. Since AB is processed prior to AC, AD, BC, and BD, there is no need to consider all six entries since the postings list that correspond to documents containing the simultaneous occurrences of A and B are already captured while AB was processed. Similarly, we need to process only CD because it will contain the list of all the documents containing C and D. Entices AC, AD, BC, and BD need not be processed. This reduces the number of postings list collected (line 10) that need to be considered while performing the intersection operation. If any keyword is not connected to the other keyword in *S* or it does not exist in the GBI then there cannot be any common document, and therefore, NULL is returned (lines 13–21). If the number of search keywords is 2 and if a nonempty postings list is collected then this list is the desired output that is returned (line 24). Otherwise, an intersection operation is performed (line 26–line 28) to determine the list of documents that contain all the keywords in S. Note that line 26 refers to Algorithm 3 for from (Mohideen et al. 2020) that is based on the technique for performing intersection for inverted indexing described in Culpepper and Moffat (2010) available in the literature. Algorithm 3 takes a hash table containing a set of postings lists and returns a NULL or an array containing the document IDs that are common among the postings lists.

9.7 Performance Analysis

A performance analysis of the GBI technique based on experiments performed with a synthetic workload is presented in Mohideen et al. (2020). The details of how the workload was generated are presented in the paper. The performance metrics used, and a summary of the important performance results are described in this section.

Performance Metrics:

Search Latency (T_S): the latency associated with BAQ execution when GBI is used. It is the average latency associated with the execution of a Boolean AND search algorithm used when the Inverted index was used for document indexing or that achieved when the GBI algorithm was used for indexing. The latency was measured as the difference between the end time and the start time for the search algorithm used in the respective cases.

Indexing Latency (T_i): is the latency associated with the indexing of a set of documents. The indexing time for any given document is the difference between the end and start times for indexing the document using GBI or Inverted Index. For a given technique, T_i is the sum of the indexing time for all the documents in the set.

Memory (M_i): is the amount of memory associated with a given indexing technique: GBI or Inverted Index. It is the memory required to store the respective hash table.

Two important workload parameters that seem to have an important impact on system performance are described.

D_{count}: the number of documents in the document set being indexed

SK_{count}: the number of keywords in the search query

Note that each Boolean AND query searches for the simultaneous existence of a given set of SK_{count} keywords in the same document.

9.7.1 Search Latency

The relationship between various parameters and search latency is presented in the following subsections.

9.7.1.1 The Impact of D_{count} on Search Latency

The effect of D_{count} on, T_s, the search latency of Boolean AND queries is captured in Figure 9.6. D_{count} is varied from 2 to 10 million and the respective values of T_s were measured on the system. Three indexing techniques are considered: GBI, Inverted Index, and Elastic Search. Since the searching complexity increases with an increase in the number of documents, T_s is observed to increase with an increase in D_{count} for all three techniques. GBI and Inverted Index were implemented by the authors of Mohideen et al. (2020) whereas Elastic Search is a well-known enterprise-level search engine that uses inverted index.

GBI demonstrates the best performance. Since the size of the postings list increases with an increase in D_{count} Inverted Index performs comparisons of more

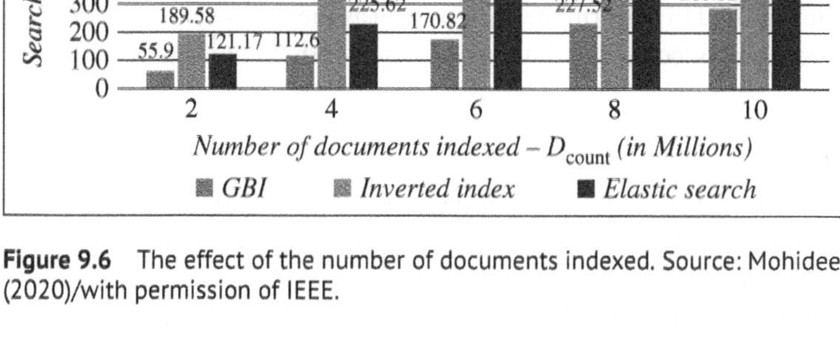

Figure 9.6 The effect of the number of documents indexed. Source: Mohideen et al. (2020)/with permission of IEEE.

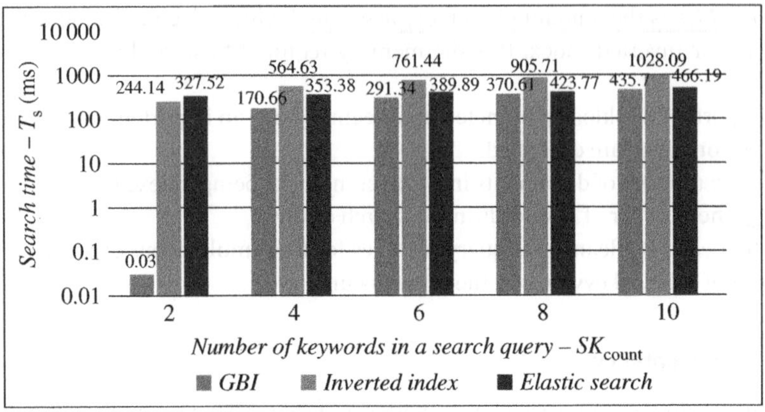

Figure 9.7 The impact of the number of keywords on search latency. Source: Mohideen et al. (2020)/with permission of IEEE.

document IDs. Although the documents list for each keyword pair increases with D_{count} also for GBI a much smaller number of intersection operations is performed for GBI, which leads to the smallest T_s for any given value of D_{count}.

9.7.1.2 Impact of SK_{count} on Search Latency

The effect of increasing the number of search keywords in the Boolean AND query can be observed in Figure 9.7. GBI once again displays the best performance. $SK_{count} = 2$ gives rise to an interesting case in which the value of T_s for GBI is very small. This is because since GBI performs indexing in terms of keyword pairs the resulting postings list can be determined by a single hash table access. For higher values of SK_{count} both GBI and Inverted Index give rise to a higher number of intersection operations to be performed. However, the number of such operations is much smaller in the case of GBI, as a result of which it displays a superior performance. For all the number of keywords experimented with, GBI demonstrates superior performance in comparison to elastic search.

9.7.2 Indexing Overhead

Both the time required to perform indexing as well as the memory used are indicators of the overhead associated with indexing. Each of these performance metrics is discussed.

9.7.2.1 Indexing Time

The time required to perform the indexing is presented in Figure 9.8, which shows the indexing time T_i for various values of the number of documents D_{count}. For any given indexing technique T_i is observed to increase with an increase in D_{count}. For any given D_{count}, GBI displays a higher indexing time in comparison to the

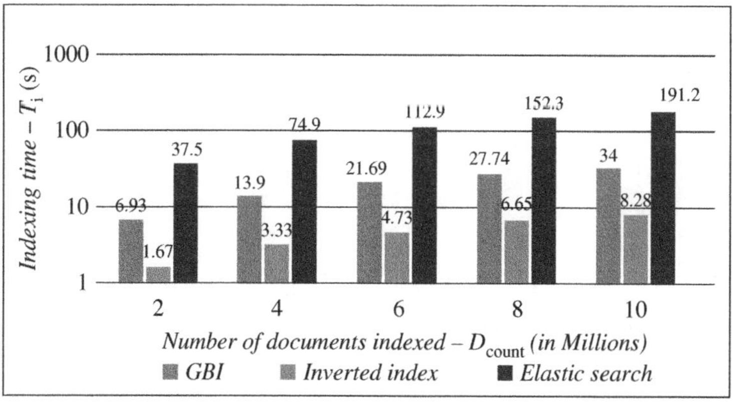

Figure 9.8 The effect of number of documents indexed on indexing time.

Inverted Index. For the inverted index, there is a postings list for each keyword. In the case of GBI, each pair of keywords gives rise to a separate postings list. Thus, if there are *n* keywords, an Inverted index will give rise to *n* postings list, whereas GBI will create *n* choose 2 postings list. Moreover, GBI needs time to create connections among the nodes in the graph. As a result, GBI requires more time to perform the indexing operation. This implies that the faster searches that accrue from using GBI come at the cost of increased indexing time. However, GBI is still considered to be beneficial because the indexing is performed only once, but the searches are performed multiple times. Thus, the indexing overhead is amortized over many search operations.

9.7.2.2 Memory Usage

The storage overhead of the three indexing techniques is presented in Figure 9.9 that shows the amount of memory M_i associated with the respective techniques.

For any given D_{count} GBI leads to more memory usage in comparison to the two other techniques. This implies that the lower search latency by GBI comes at the cost of using more memory on the system.

OR an NOT queries can also be handled by GBI-based techniques. The interested reader is referred to Mohideen (2020) for a discussion of these techniques.

9.8 Data Filtering

The reduction of search latency was discussed in Section 9.3. As discussed in Section 9.2, another method of speeding up data processing is through the reduction of data volume by performing data filtering. This will be the subject of discussion for the subsequent sections of this chapter.

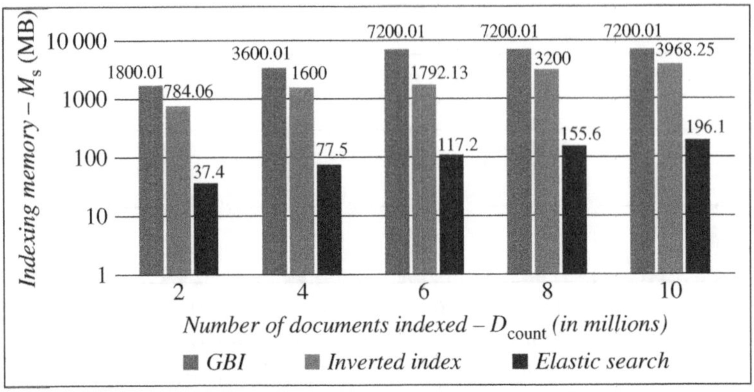

Figure 9.9 The effect of number of documents indexed on memory used.

9.8.1 Processing of Large Volumes of Data

Processing of data can be performed in various ways. Batch processing and stream processing are two different techniques for processing large volumes of data (see Figure 9.10). Batch processing is performed with data at rest, and it thus deals with stored data files. One of the challenges for batch processing systems is to handle large volumes of data. Both the volume and velocity characteristics are relevant in the context of stream processing systems since they need to handle large volumes of streaming data in a timely manner. One of the challenges of the stream processing techniques is the need to analyze data quickly and, in some cases, generate the results of data analysis within a specified interval of time.

Figure 9.10 lists the four challenges of big data (volume, velocity, variety, and veracity) introduced earlier in the chapter. The current discussion focuses on addressing the volume and velocity challenges. Data can be generated from various sources such as smartphones, desktop computers, smartwatches as well as social networks. This data are then ingested by the system first before it is sent for batch or stream processing, depending on the application needs.

Parallel processing platforms comprising multiple compute elements and disk drives are often used for addressing the large compute and storage requirements of the applications.

9.9 Parallel Processing Platforms

The MapReduce/Hadoop platform that is often used in analyzing stored data are discussed in Chapter 6 of this book. This chapter discusses Apache Spark,

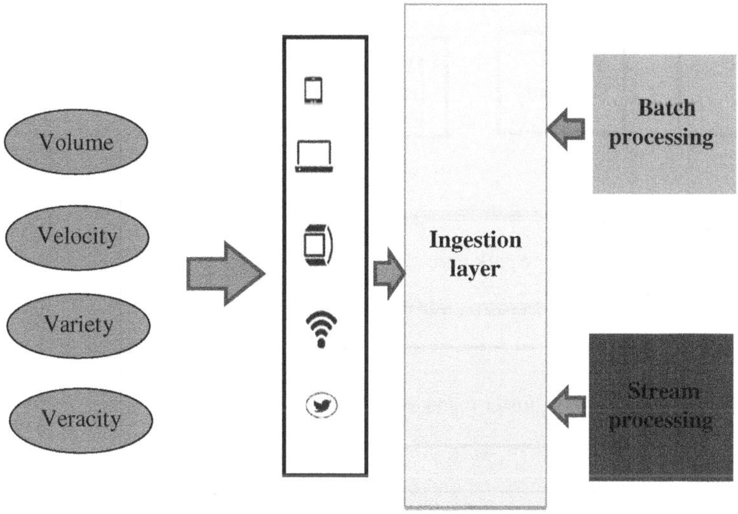

Figure 9.10 Two forms of data processing.

a parallel processing framework that is useful for processing large data sets. It addresses a performance shortcoming of its predecessor Apache Hadoop in which the intermediate results of an operation performed on a MapReduce program, for example, needs to be written onto disks, thus slowing down the system. Apache Spark alleviates this problem by using only in-memory computations, thus avoiding the use of slower electro-mechanical devices such as disk drives for transferring data between the map and reduce tasks. The application areas for Spark include e-commerce, social media, entertainment, and health care. Examples of companies that use Spark include eBay and Netflix, which are described in Kalcheva and Nikolov (2020).

As indicated earlier, Spark processes large data sets by using a parallel processing framework. It provides an easy-to-use API that enables the user to utilize the underlying parallel computing platform achieved with the incorporation of multiple computing nodes. It is a hybrid system because it provides support for both batch and stream processing. The Apache Spark ecosystem is shown in Figure 9.11. It runs on existing environments such as Hadoop, Apache Mesos, and Kubernetes, as well as on a cloud with its own scheduler. It provides APIs for a number of high-level languages, such as Java, Scala (Scala n.d.), Python, and R (R n.d.).

Spark interacts with disk systems only at the beginning for loading the input data and at the end for writing the results of the computations performed. All intermediate results are stored and processed in memory with the help of Resilient distributed datasets (RDDs) (Karau and Warren 2017), an important component

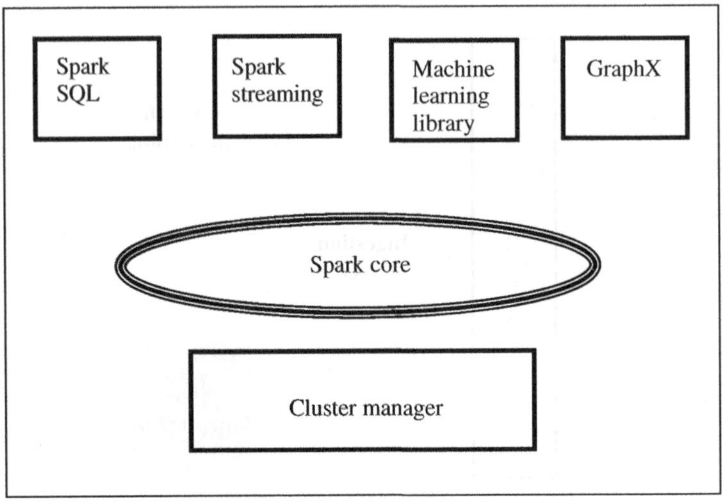

Figure 9.11 Key components of the spark ecosystem.

of the Spark Core. An RDD is a read-only data structure that is memory resident and is used to represent large datasets. RDDs are distributed collections of objects often referred to as partitions (Karau and Warren 2017). They are used in conjunction with two types of operations: *transformations* and *actions*. A transformation creates a new RDD based on an existing RDD but does not return any value. An action, on the contrary, returns a value without creating a new RDD. An example of a transformation is a "filter" that returns a new RDD based on the filter condition (Apache Spark n.d.-a). An example of a spark action is "collect," which returns the contents of an RDD to the driver node in the Spark cluster (Apache Spark n.d.-a).

The Spark system utilized in data analytics applications uses a number of interconnected components. The key components of this ecosystem depicted in Figure 9.11 are briefly described.

Spark SQL (Apache Spark n.d.-b): provides the capability of structured data processing. It provides the DataFrame and DataSet abstractions that enable Spark to execute Structured Query Language (SQL) queries. A DataSet is a distributed collection of data, and a DataFrame is a distributed set of data organized into columns. It is useful in the context of analyzing both stored as well as streaming data.

Spark Streaming: is a high throughput stream processing component of Spark. It is a stream processing engine built on top of Spark SQL. It is useful in processing streaming data generated in real time. Examples include sensor data, logging data, and status updates posted by users on social networks.

Spark MLib: is the machine learning library of Spark containing common machine learning algorithms and tools.

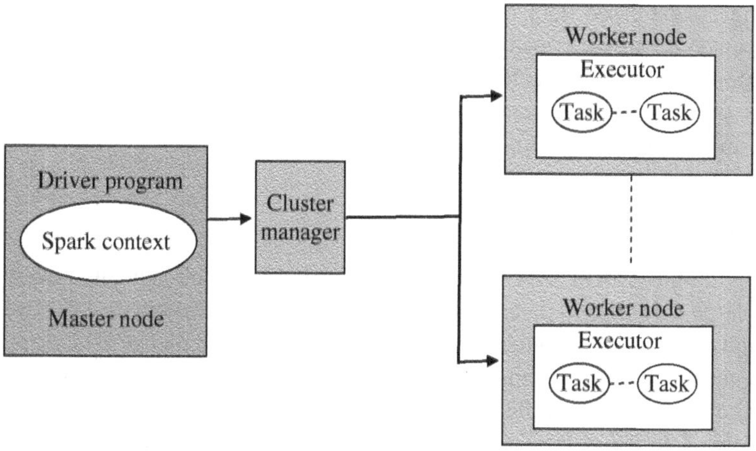

Figure 9.12 Architecture of spark.

GraphX: is a framework provided by Spark for graph processing. GraphX provides APIs for executing parallel algorithms.

Spark Core: provides support for RDDs and programming languages such as Python, Java, SQL, and R.

9.9.1 Spark Architecture

A master–slave architecture is used in Spark. The Master Node runs the Driver Program whereas as the slave nodes referred to as Worker Nodes can operate in parallel and are used to execute tasks (see Figure 9.12). The Driver Program is the main coordinator of Apache Spark. It calls the main program in a given application and creates a Spark Context that is the main entry point for all Spark functionalities. It translates the application into jobs, each of which is further split into tasks that run on the Worker Nodes. The Driver program is also responsible for scheduling job execution and negotiating with the Cluster Manager for resource allocation. Various types of Cluster Managers including YARN (Apache Hadoop n.d.) and Mesos (Apache Mesos n.d.) can be used.

The Spark system will be used in the discussions presented in the upcoming sections.

9.10 Motivations for Data Reduction

The following sections concern the volume characteristics of big data and discuss techniques for handling large volumes of textual data for specific users.

Processing of text-based data needs to be performed for various reasons that include personal, business, medical, and academic purposes. Users often rely on using large textual documents that can contain meeting minutes, tweets as well as articles published in conferences and journals for making important decisions (Chanda and Majumdar 2021a).

9.10.1 Reducing Data Volume

According to Marr (2018), over 2.5 quintillion bytes of data are generated every day. The large volume of data are indeed one of its biggest challenges. Effective storage and processing techniques are needed to address this challenge of big data. The data generated every day may span various topics. The data on a given topic itself can be very large. A specific user, however, may be interested in only a small subset of this data. Thus, if the original large raw data can be reduced to contain only the data this user is interested in the reduced volume of data can be stored and processed more efficiently. A set of example uses cases in which data reduction is useful is presented.

The meeting minutes use case: Companies often store minutes of meetings that keep on accumulating over time. A specific employee in the company, however, may be interested in some of the topics discussed during these meetings collected over a given period of time (e.g. years, months, and days). For example, they may be interested in meeting minutes that contain discussions on specific products and specific employees. This can form a much smaller size of data that this employee needs to store for their future use.

The medical practitioner use case: Voluminous medical records are stored in hospitals and clinics. A given doctor may be interested in only a small subset of the patient files. These may be files of patients characterized for example by their ailment types, last names, age, the names of the specific clinics they visited and a combination of one or more of these. Clearly, presenting the specific doctor with the information on such selected patients can reduce the data they need to process, thereby significantly reducing the data processing time.

The journalist use case: A large number of tweets (e.g. 500 million Sayce n.d.) are produced every day. A specific journalist is often interested in tweets that contain specific key phrases or words. Distilling the total number of tweets to retain only those that are of interest to the specific journalist can effectively increase their productivity.

The academic researcher use case: A very large number of articles is published regularly in journals and conference proceedings in any given field of Science and Engineering. A specific researcher is typically interested in only a small subset of these papers in their chosen fields and topics of interest. Reducing this large set of papers published to a small subset of papers the researcher is interested in can be very useful for the researcher.

User preferences specified by a user can be used to perform this filtering (see Section 9.11.1 for details). In some situations, it may also be possible to derive user preferences by using the user's past behavior in searching for topics of interest (Zhang and Pu 2007) or the similarity of a new user with other similar existing users (Su et al. 2018).

9.11 Data Filtering

An example of how user preferences are used to filter data are provided.

Consider the raw data comprising the following three sentences.

Sentence 1: "Your dental appointment with Dr. Henry is scheduled at 10:00 a.m. on February 16."

Sentence 2: "Please meet the dental clinic staff five minutes before your appointment."

Sentence 3: "The location of the dental clinic is provided in the following URL: http://www.Dentalclinic.ca."

The user preferences contain "person" and "date." Sentence 1 contains both the name of a person and a date, whereas Sentence 2 refers to a person "dental clinic staff." Sentence 3, on the other hand, does not refer to any person or date. As a result, Sentence 1 and Sentence 2 (and not Sentence 3) will be included in the filtered data.

9.11.1 Basic Approach for Data Filtering

The data filtering approach for reduction of data volume is explained with the help of Figure 9.13 that describes the filtering operation performed on a cloud (Chanda and Majumdar 2021a).

The raw data set is the data that needs to be filtered based on user preferences. It comprises one or more text data files. As shown in Figure 9.13 the raw data set is processed by the filter method that applies user preferences for data filtering. The user provides the filter method with this set of preferences that indicates the topics from where they will be performing searches in the future. The preferences are in the form of a set of keywords where each of the keywords describes a topic the user is interested in, such as name, date, and place. The filter module searches through the raw data and identifies the components that match one or more user preferences. These matched components are then saved as filtered data, whereas the remaining raw data are discarded. Once filtering is complete the user can

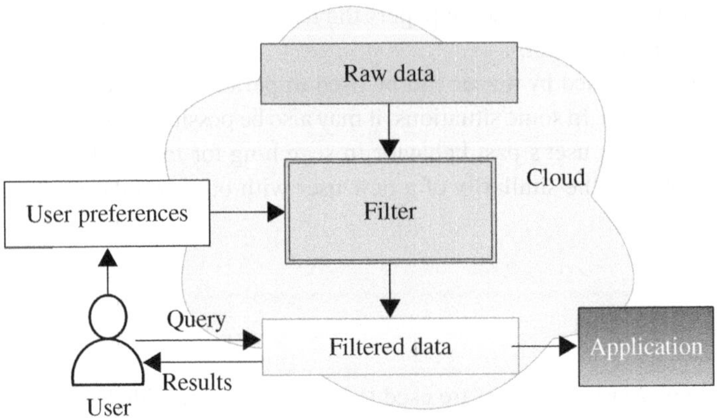

Figure 9.13 High-level approach for data filtering. Source: Chanda and Majumdar (2021a)/with permission of IEEE.

use the filtered data as many times as they like and thus the overhead of filtering is amortized over many searches. The user performs a search for a specific data item by sending a search query. In addition to responding to search queries, the filtered data may be used by specific applications. An example is an application that goes through the raw data that is filtered resulting in data that contains dates and names. The application uses this filtered data for generating a meeting schedule for the user for a given range of days.

The important components of the filter module are described next (see Figure 9.14)

The raw data are presented as one or more text files to the filter method. Each step captured in Figure 9.14 is described.

Text Preprocessing: In this step the raw data are prepared for the filtering operation. The step includes eliminating punctuations and lowercasing the uppercase letters. This data can then be analyzed by the Named Entities Recognition (NER) function and the machine learning model that is discussed next.

Text Categorization: refers to the classification of the preprocessed text with the help of the NER function and a Machine Learning model. The NER function provided by the Python library classifies the raw text data and extracts specific entities such as people's names, locations, dates, and money from the raw data. For the Machine Learning model, the technique described in Chanda and Majumdar (2021a) uses a Multinomial Logistic Regression Classifier of Apache Spark's machine learning library. The classifier runs a sequence of pipeline stages that are presented next.

Tokenization: is the process of reading the input text and dividing it into a number of meaningful entities such as words. Of particular interest are Stop words such

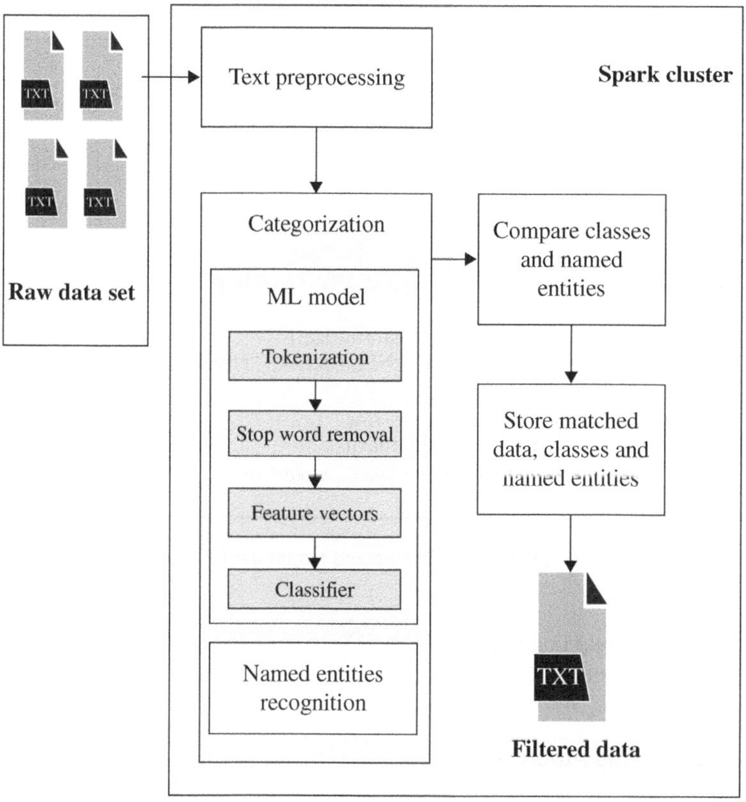

Figure 9.14 Key components of the filter module. Source: Chanda and Majumdar (2021a)/with permission of IEEE.

as "the," "an" and "a" that do not often contribute directly to the subject matter discussed.

Stop Word Removal: is the process of removing stop words from tokenized data. In Chanda and Majumdar (2021a), for example, Apache Spark's MLib "StopWordRemover" class that takes the tokenized text and removes the stop words is used.

Feature Vector: Feature extraction is performed in the next step to produce the feature vectors. This step derives numerical features from the raw text data that can then be used by machine learning algorithms (Kumar and Rajini 2019). Each component of the raw text data are mapped into a vector space that can be utilized in the training of the machine learning model (Nabi n.d.). This mapping from the text data space to the real valued vector space is performed during feature extraction. The output of the feature extraction step is the feature vector that forms the

Pseudo Code 1: Filter Method

Input: User preferences, P[] and raw dataset, R
Output: User preferred information

```
1: srdd ← read the raw dataset R.
2: data ← srdd.map {
3:    line ← sent_tokenize(line),
4:    line ← preprocessing(line),
5:    line ← NER(line, P)
6:    Row(Text=line[0], NERTokens=line[1], NERData=line[2])
7: }.toDF()
8: prediction ← MLPipelineModel.transform(data)
9: selected ← prediction.select("Text", "NERTokens", "NERData", "label")
10: for row in selected.collect():
11:        text, nertokens, nerdata, label ← row
12:        if compareClasses( P, label, nertokens):
13:              SavetoFile(text, label, nertokens, nerdata)
```

Figure 9.15 Filtering algorithm. Source: Chanda and Majumdar (2021a)/with permission of IEEE.

input to the Multinomial Logistic Regression Classifier used in data classification by the technique described in the paper (Chanda and Majumdar 2021a).

9.11.2 The Filtering Algorithm

The filtering techniques deployed on a Spark cluster are described with the help of Figure 9.15 (Chanda and Majumdar 2021a) that presents the pseudo code for the algorithm.

The input to the algorithm is (i) a directory where all the raw data files are located and (ii) user preferences. First, the Filter method reads the raw dataset from the specified raw data set directory R, and creates RDDs by using a Spark system call (line 1). Note that the raw data set is distributed among the executor cores that operate in parallel. Then, the text data from each raw data file are processed using the Spark "map" function (lines 2–6). At the beginning (line 3), the text data elements are partitioned into a set of sentences and each sentence is preprocessed in line 4. In this preprocessing step, the uppercase letters are converted into lowercase and the punctuations in the sentence are removed.

The NER function is then applied to each sentence for extracting the named entities such as people's names, products, and dates in accordance with user preferences contained in P (line 5). In line 6, the outputs of NER are used to create row instances, where each row comprises three columns: "Text," "NERTokens," and "NERData" that correspond, respectively, to the sentence, named entities classes, and the extracted named entities. The DataFrame with the three columns is created in line 7 and then passed as input to the machine learning model to classify the sentences according to their content (line 8). The model returns the label for each class as an additional column "label" for the predicted class. The algorithm then iterates through each row to compare (lines 10–13) the predicted classes with user preferences in P with the help of the function compareClasses. If there is a match between the predicted classes for a given row and the user preferences, the contents of the row are saved in the filtered data (line 13). The filtered data are stored in a csv format.

9.11.3 Search Method

Once the filtering of raw data are complete, the user can search for specific data items in the filtered data. The search is performed by using a linear search algorithm. User queries can be sent to the search method in two different ways. In the first, the user sends keywords to the search method (see Figure 9.16a), and in the second, the user sends a sentence to the search methods (see Figure 9.16b).

For searching by keywords, the search method performs text processing to convert the data in csv format available in the filtered data repository into sentences and divides the resulting data equally among all the executor cores. Each core searches for the occurrence of the keyword in the data partition allocated to it, and the search results obtained from all the cores are returned to the user.

The steps associated with searching by sentences (see Figure 9.16b) are discussed. The query sentence is preprocessed during the Query cleaning operation that eliminates punctuation marks and converts uppercase letters to lower case letters. By using the same machine learning models and the NER function discussed in the context of data filtering (see Section 9.11.2) the sentence is classified and the named entities are extracted from the sentence. A method similar to the text processing function described in the context of searching by keywords is used to process the filtered data available in the csv format. A list of sentences, their classes and named entities are obtained at the end of this text-processing step. These are then compared with the classes and named entities in the user provided sentence. The search method then returns those sentences from the filtered data whose classes and named entities match with the ones obtained at the end of the Categorization operation (see Figure 9.16b).

Consider the example discussed in Section 9.11. Recall that the filtered data comprises the two following sentences.

Sentence 1: "Your dental appointment with Dr. Henry is scheduled at 10:00 AM on February 16."

Sentence 2: "Please meet the dental clinic staff five minutes before your appointment."

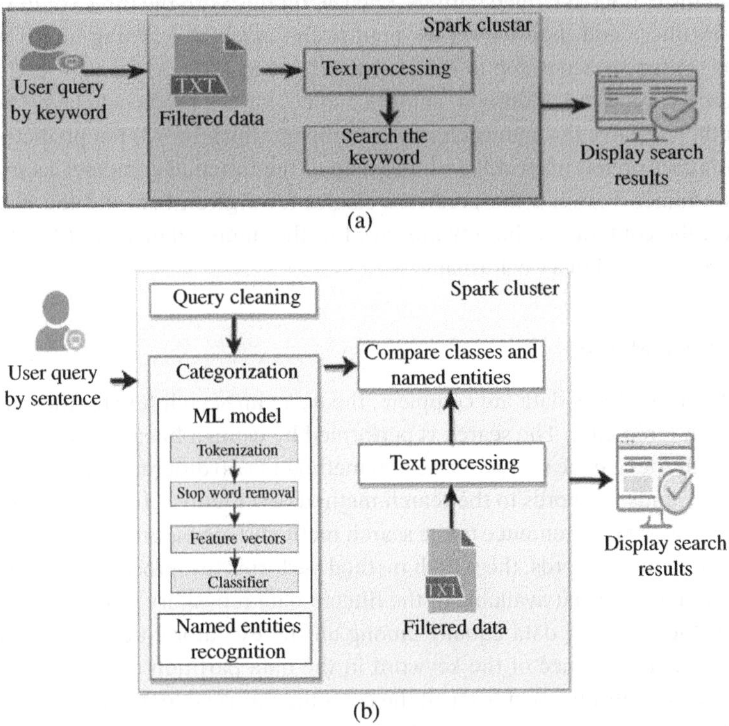

Figure 9.16 The two methods of sending search queries: (a) searching by keywords and (b) searching by sentences. Source: Chanda and Majumdar (2021a)/with permission of IEEE.

If the user sends the user queries by a sentence such as "When is my appointment with Dr. Henry?" the search method will preprocess and classify the query as "Person, Date" and extract the named entity "Henry." The class and named entity in the filtered data are then searched, and the search method displays sentence 1: "Your dental appointment with Dr. Henry is scheduled at 10:00 a.m. on February 16."

9.12 Performance Analysis

The performance of the filter method is analyzed by using synthetic raw data sets that comprise replicated meeting minutes (see Chanda and Majumdar 2021a for details). The performance metrics used are described next.

- *Computation time* (T_C)*:* is the time taken by the filter method to filter a given raw dataset. Given a raw data set and the number of worker nodes, N_W is measured

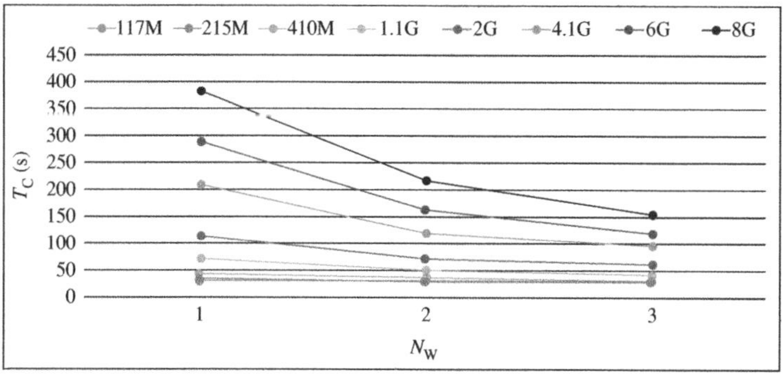

Figure 9.17 The impact of the number of worker nodes. Source: Chanda and Majumdar (2021a)/with permission of IEEE.

as the difference between the time at which the filter method completes and the time at which it starts execution.

- *Speedup (S(N)):* is the ratio of T_c achieved on a single core and that achieved with N cores of worker nodes. It estimates how well the application speeds up when the parallelism provided by multiple CPU cores increases.

Note that the worker nodes have multiple cores, and N is the total number of all the cores across all the worker nodes.

- *Efficiency (E(N)):* for a given N is the proportion of time for which the N processing elements are usefully utilized. It is computed as the ratio of T_c achieved on a single core and the product of T_c achieved with N cores and N. Recall the relationship between $S(N)$ and $E(N)$ presented earlier in the book. Thus, $E(N)$ can also be computed as the ratio between $S(N)$ and N.

The impact of the number of worker nodes and raw data size on filtering time is presented in Figure 9.17 (Chanda and Majumdar 2021a). Each of the N_w nodes was equipped with an eight-core CPU. For a given raw data size, as the number of worker nodes, N_w, increases, the parallelism of execution increases. Since more data can be processed with an increase in the number of worker nodes, T_c tends to decrease as N_w is increased. The decrease in T_c with an increase in N_w is sharper for large sizes of raw data.

The relationship between $S(N)$, $E(N)$, and N is captured in Figure 9.18. As the number of worker node cores N is increased parallelism in execution increases and speedup tends to increase with N. Since all the cores are not fully utilized for 100% of the time, $E(N)$ is observed to decrease as N is increased.

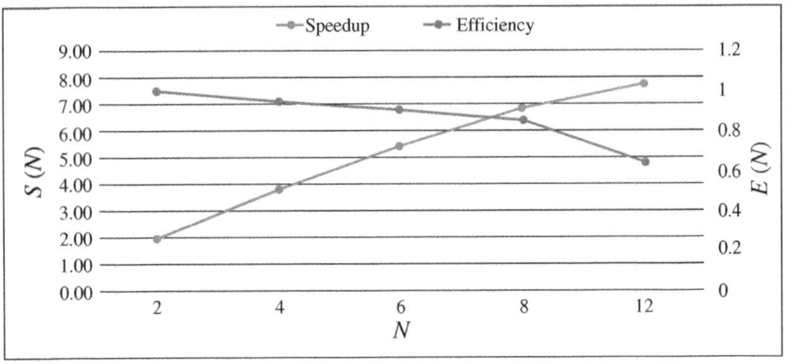

Figure 9.18 Relationship between speedup, efficiency, and *N*. Source: Chanda and Majumdar (2021a)/with permission of IEEE.

9.12.1 Node Parallelism vs. Core Parallelism

Concurrency in execution can be increased in two ways: by increasing the number of worker nodes or by keeping the number of worker nodes at a low value but increasing the number of cores per worker node. Which of these options will give rise to a higher performance? This question is answered by conducting an experiment in which the total number of worker node cores is fixed at 12, but the number of worker nodes N_w is varied from 1 to 3. Note that the number of cores per worker node decreases from 12 to 6 to 4 As N_w is increased from 1 to 2 to 3.

The results are presented in Figure 9.19 (Chanda and Majumdar 2021a). T_c is observed to decrease with an increase in N_w. Thus, using a lower number of cores per worker node seems to be preferable. This is because with multiple cores in one worker node, the overhead of collecting the results from all the cores increases as the number of cores in the node increases.

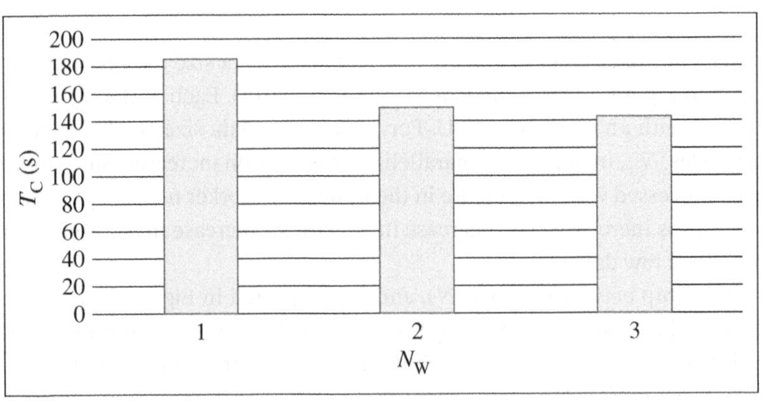

Figure 9.19 Node parallelism and core parallelism. Source: Chanda and Majumdar (2021a)/with permission of IEEE.

9.13 Streaming Data

As in the case of filtering data at rest, users are often interested in filtering data from raw data streaming in real time. Examples include tweets and sensor data in internet-of-things-based systems. The parallel processing framework such as Spark discussed earlier in this chapter is often used for processing streaming data. Spark receives streaming data through a middleware that includes publish subscribe systems. Data comes from a data source often referred to as a publisher that sends the data to the publish subscribe middleware. Once or more receivers also referred to as subscribers or consumers can receive the data that it is interested in from the middleware. A subscriber subscribes to one or more topics that describe the classes of data that it is interested in. Only the data that belongs to a specific topic subscribed to by a given subscriber is delivered to it.

An example of such a publish subscribe system is Apache Kafka (Estrada 2018) that receives data from publishers and brings the data to the streaming engine of Spark. The incoming data are partitioned across multiple topics. Topics get placed into individual partitions by Kafka.

The filtering method used for filtering streaming data are explained with the help of Figure 9.20 (Chanda and Majumdar 2021b). The input to the filter method is provided by the batches of records stored as Kafka topics. This data are divided equally among the executor cores.

The executor cores run in parallel, with each core handling its own partition of data. During text preprocessing the punctuation marks are removed and the uppercase letters are converted into lowercase. During Text Categorization, the preprocessed data are classified by using NER and the Machine learning model that was discussed in the context of filtering stored data in Section 9.11.2. The NER function extracts entities such as person names, locations, dates, and numbers. A multinomial logistic regression classifier is used by the machine learning model

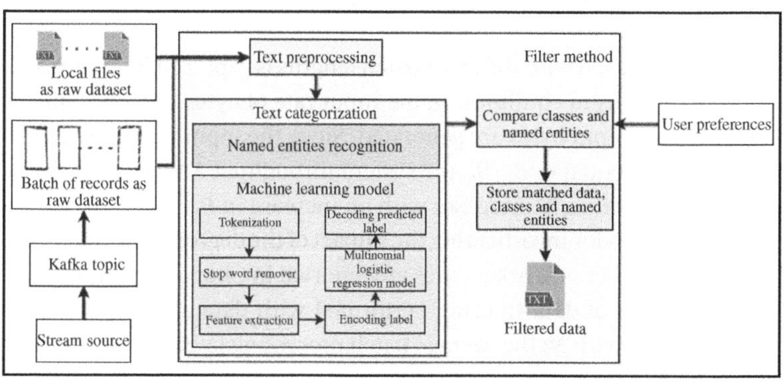

Figure 9.20 Filtering of streaming data. Source: Chanda and Majumdar (2021b)/with permission of IEEE.

to classify text based on its contents. The specific classes of the classified text are compared with the user preferences and the matched data are stored as filtered data in the form of a csv file.

9.13.1 System Performance

An investigation of the performance of data filtering for streaming data are reported in Chanda and Majumdar (2021b). A synthetic raw data set is generated by using several medical conversations between patients and doctors reported on two websites. A producer application in the workload generator reads this raw data file and sends batches of records to Kafka, which sends the data to the filter method (see Figure 9.20). The filter method hosted on Spark processes the data in parallel.

The key parameters of the streaming workload are discussed.

Batch Length (B_L): is the number records in a batch. Note that each record comprises a single sentence.

Batch Interval (B_I): is the time between the sending of successive batches. Note that after sending a batch, the producer in the workload generator waits for an interval of time B_I before sending the next batch so that the filter method can complete the processing of a batch before the next batch arrives.

The performance metrics discussed in this section are introduced next.

Average Batch Processing Latency (T_p): is the average time taken by the filter method to process a batch of records.

Throughput (X): the total number of bytes processed by the filter method per unit time.

The results of an experiment in which the batch interval, B_I, and the number of worker cores, N, were fixed, and the length of the batch was varied is presented in Figure 9.21.

Each record in the raw data needs to be compared with the user preferences. Thus, increasing the number of records in a batch increases the average time to process a batch, T_p.

The throughput achieved with different batch lengths is captured in Figure 9.22. Since the system remains in equilibrium, the input rate of bytes must be equal to the rate at which the output bytes are generated. Since the input rate of bytes that flow into the system is equal to B_L/B_I, the system throughput X is equal to B_L/B_I. As B_I is held at a fixed value, X increases with an increase in B_L.

Results of experiments for investigating the impact of the degree of parallelism as reflected by N, the number of Worker cores are reported in Chanda and Majumdar (2021b). As in the case of data filtering performed with data at rest the system performance improves with N: the average batch processing latency is observed to decrease with an increase in N.

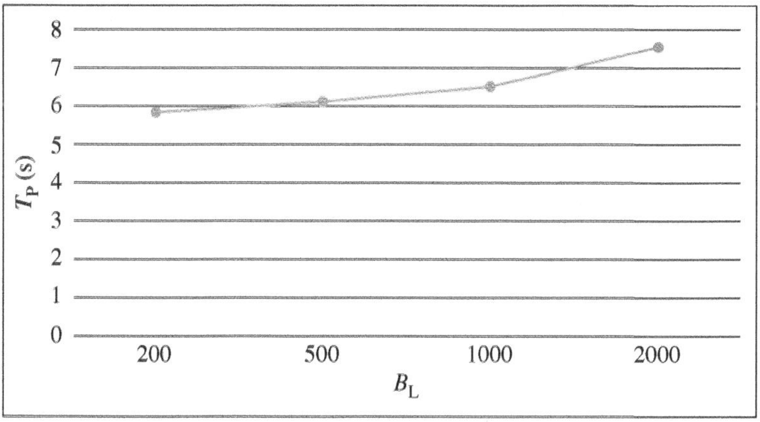

Figure 9.21 The impact of batch length on batch-processing latency. Source: Chanda and Majumdar (2021b)/with permission of IEEE.

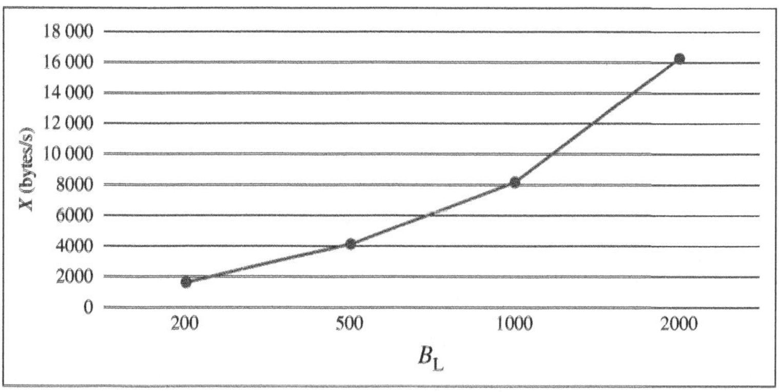

Figure 9.22 The impact of batch length on throughput. Source: Chanda and Majumdar (2021b)/with permission of IEEE.

Filtering Efficiency: One of the benefits of data filtering is the reduction in search latency. This benefit is captured in filtering Efficiency E_F, which is the ratio of the search latency observed for a given argument for the search method achieved with the search performed on raw data and that achieved with filtered data. Sample test results reported in Chanda and Majumdar (2021b) are presented in Table 9.3. The search algorithm is implemented as both a sequential and a parallel program. The type of the search method used (sequential/parallel) is presented in the first column. The second column presents the type of the search argument (single keyword or sentence), whereas the third column presents the value of E_F. Search efficiency varies from 57.6 to 105 for sequential searches and from 29 to 63.6 for

Table 9.3 Filtering efficiency.

Search method	Search by	E_F
Sequential	Keywords	105
Sequential	Sentences	57.6
Parallel	Keywords	63.6
Parallel	Sentences	29

Source: Chanda and Majumdar (2021a)/with permission of IEEE.

parallel searches. For both types of search methods, search by keyword gives rise to a higher E_F in comparison to search by sentence.

9.14 Handling User Preferences Comprising Keywords Connected by Boolean Operators

A user-specified keyword can sometimes comprise multiple keywords/sentences connected by Boolean operators such as AND, OR, and NOT (Davidson and Majumdar 2022). An example of such a use case is presented next.

KEYWORD: ("lemons" AND "oranges") OR "human"

RAW DATA:

1. Both lemons and oranges are sources of Vitamin C.
2. Getting enough Vitamin C daily is good for the human body.
3. Canada imports oranges from the USA.

In this example, the user is interested in saving sentences that either contain the keywords lemons and oranges or the word human. The filtered data, in this case, will thus contain sentence 1 that matches ("lemons" AND "oranges") and sentence 2 that includes "human."

The filter method used in (Davidson and Majumdar 2022) uses SQL operations during filtering to find matches between keywords containing logical operators and raw data.

9.14.1 Performance Analysis

The impact of the type of keywords (Simple and Complex) and the number of operators used in connecting multiple keywords is discussed in (Davidson and Majumdar 2022). A Simple keyword uses only a set of N_f keywords connected by Boolean operators: OR or AND whereas a Complex keyword comprises N_f sets of keywords each of which consists of three words that are connected by Boolean operators: OR or AND. The performance metric of interest is the Logical Query

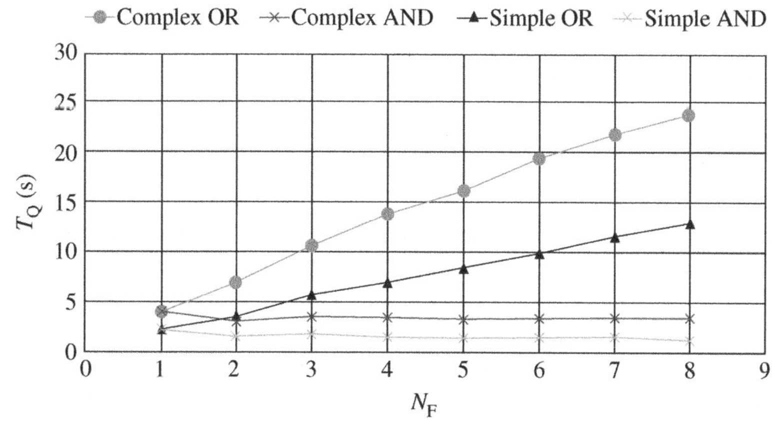

Figure 9.23 The effect of the number of filter terms on performance. Source: From Davidson and Majumdar (2022).

Time (T_Q), which is the time taken for processing the SQL query used in filtering when user preferences use Simple or Complex keywords. It is measured as the difference between the time at which results of the SQL operation in the Filter method are collected on the Spark's Master Node and the time at which the SQL operation in the Filter method is initiated by the Spark's Master Node.

Figure 9.23 presents the results of an experiment described in (Davidson and Majumdar 2022). The number and type of keywords were varied, and the corresponding T_Q was measured. Multiple classes of Boolean expressions used for combining keywords were experimented with. Each class included two categories of expressions: "OR" structured expressions consisting of the logical "OR" of multiple keywords and "AND" structured expressions comprising the logical "AND" of multiple keywords. The results of the experiments for both simple and complex types of keywords are presented in the figure. The type of the keyword is captured in the label of each line. N_f is the number of filter terms used in a given experiment. Complex OR and Complex AND give rise to a higher T_Q in comparison to their Simple counterparts. This is because of the higher number of logical operations required for the Complex keywords. For a given type of OR expression (Simple or Complex), the logical query time T_Q increases with the number of terms N_f. This is because a higher value of N_f leads to a higher number of matches thus increasing both the pattern-matching time as well as the result collection time. For both Simple and Complex AND, however, the increase in N_f does not seem to change the value of T_Q. Note that the probability of finding a mismatch in any one term increases with N_f. and a mismatch in any one of the AND operations results in a mismatch for the entire expression evaluation. As a result although the pattern-matching time increases with an increase in N_f, the results collection time decreases. The decrease in results collection time seems to offset the increase in pattern matching time, thus resulting in almost the same value of T_Q for different values of N_f.

9.15 Summary

This chapter focused on the challenges of big data. It described a number of important characteristics of big data: volume, velocity, variety, and veracity that make the analysis of big data difficult. Two different approaches for addressing these challenges were discussed: decreasing search latency and reducing data volume. Data indexing is a well-known method for improving search latency. A number of data indexing techniques were discussed. Speeding up processing through the use of the MapReduce parallel processing framework was discussed in Chapter 6 of the book. This chapter introduced another parallel processing framework, Spark that addresses some of the limitations of MapReduce by replacing the saving of intermediate data on disks by in-memory processing.

Different data filtering techniques that can reduce the volume of raw data were discussed. Techniques for handling both data at rest (stored data) and data in motion (streaming data) were described. The techniques filter raw data by using user-provided keywords/sentences. Filtering is a time-consuming operation and deployment of the techniques on Apache Spark helps reduce the data filtering time. The performance of the system for different combinations of systems and workload parameters was discussed. Insights into system behavior and performance that accrue from the performance analysis were reported.

Exercises

1 Data indexing and data filtering are two different approaches for reducing data search time. Compare the two approaches in the context of the reduction of data search times as well as in terms of the overhead associated with each.

2 Consider the following sets of documents:

Document id	Keywords
1	Canada, India, China
2	United States
3	Russia, Ukraine
4	India, Ukraine, United States
5	Canada, Ukraine, Russia

Draw the graph achieved by using the GBI technique.

3 Consider the document set described in question 2. Show the mapping between index and postings list for the document set as captured by the inverted index technique.

4 Compare the inverted indexing technique with the GBI technique, showing their relative merits and demerits.

5 Distinguish between node parallelism and core parallelism in the context of data filtering. Discuss the possible rationale behind the difference in their impacts on system performance.

6 Provide the rationale behind the following observations:
(a) The decrease in T_c with an increase in N_w is sharper for large sizes of raw data (see Figure 9.17).
(b) The insensitivity of $T_Q(S)$ to N_F for both simple and complex ANDs in Figure 9.23. Provide an example to illustrate your point.

References

Analytics Insight, "*The Four V's of Big Data – What is Big Data?*", Analytics Insight, December 2021, https://www.analyticsinsight.net/the-four-vs-of-big-data-what-is-big-data/#:~:text=Big%20data%20is%20now %20generally,velocity%2C%20variety%2C%20and%20veracity. [accessed: 23-Jan-2021].

Apache Hadoop, "Apache Hadoop YARN", n.d., https://hadoop.apache.org/docs/current/hadoop-yarn/hadoop-yarn-site/YARN.html. [accessed: 23-Jan-2021]

Apache Mesos, "Apache Mesos", n.d., https://mesos.apache.org/. [accessed: 23-Jan-2021]. SM- n.d. is OK

Apache Software Foundation, "Apache Lucene," n.d., https: //lucene.apache.org/. [accessed: 25-October-2019].

Apache Spark, "RDD Programming Guide", n.d.-a, http://spark.apache.org/docs/3.0.1/rdd-programming-guide.html. [accessed: 12-Jan-2021].

Apache Spark, "Spark SQL, DataFrames and Datasets Guide", n.d.-b, Fhttps://spark.apache.org/docs/3.0.1/sql-programming-guide.html. [accessed: 02-Mar-2021].

B. Chanda, S. Majumdar, "A Parallel Processing Technique for Extracting and Storing User Specified Data", *Proc. IEEE 8th International Conference on Future Internet of Things and Cloud (FiCloud)*, Rome, August 2021a.

B. Chanda, S. Majumdar, "A Technique for Extracting User Specified Information from Streaming Data", *Proc. International Symposium on Performance Evaluation of Computer and Telecommunication Systems (SPECTS)*, Online, July 2021b.

J. S. Culpepper, A. Moffat, "Efficient Set Intersection for Inverted Indexing," *ACM Transactions on Information Systems*, Vol. 29, Issue 1, 2010, pp. 1:1–1:25.

G. Davidson, S. Majumdar, "Boolean Logical Operator Driven Selective Data Filtering for Large Datasets", *Proc. Annual Modeling and Simulation Conference (ANNSIM 2022)*, San Diego, USA, July 2022.

R. Estrada, *Apache Kafka Quick Start Guide*, Packt, UK, 2018.

M. Fontoura, M. Gurevich, V. Josifovski, S. Vassilvitskii, "Efficiently Encoding Term Co-occurrences in Inverted Indexes", *Proc. 20th ACM International Conference on Information and Knowledge Management,* 2011, pp. 307–316.

N. Kalcheva, N. Nikolov, "Laplace Naive Bayes Classifier in the Classification of Text in Machine Learning", *Proc. International Conference on Biomedical Innovations and Applications (BIA)*, 2020, pp. 17-19.

K. Kambatla, G. Kollias, V. Kumar, A. Grama, "Trends in Big Data Analytics", *Journal of Parallel and Distributed Computing*, Vol. 74, Issue 7 July 2014, pp. 2561–2573.

H. Karau, R. Warren, *High Performance Spark: Best Practices for Scaling and Optimizing Apache Spark*, O'Reilly, Beijing, 2017.

S.S. Kumar, A. Rajini, "Extensive Survey on Feature Extraction and Feature Selection Techniques for Sentiment Classification in Social Media", *10th International Conference on Computing, Communication and Networking Technologies (ICCCNT)*, 2019, pp. 1–6.

A. Kumar, S. Johari, S. Saha, "Reduction in Searching Time of Inverted Index Using Bloom Filter", *Proc. International Conference on Emerging Research in Computing, Information, Communication and Applications*, 2014.

A.K. Mahapatra, S. Biswas, "Inverted Indexes: Types and Techniques", *International Journal of Computer Science Issues,* Vol. 8, Issue 1, July 2011, pp. 384–392.

B. Marr, "How Much Data Do We Create Every Day? The Mind-Blowing Stats Everyone Should Read", *Forbes*, 2018, https://www.forbes.com/sites/bernardmarr/2018/05/21/how-much-data-do-we-create-every-day-the-mind-blowing-stats-everyone-should-read/?sh=754f85e260ba. [accessed: 18-May-2023].

T. Mikolov, K. Chen, G. Corrado, J. Dean, "Efficient Estimation of Word Representations in Vector Space," arXiv preprint arXiv:1301.3781, 2013.

A.K. Mohideen, "A Graph-Based Indexing Technique for Efficient Searching in Large Scale Textual Documents", M.A.Sc. Thesis, Department of Systems and Computer Engineering, Carleton University, Ottawa, Canada, 2020.

A.K. Mohideen, S. Majumdar, M. St-Hilaire, A. El-Haraki, "A Data Indexing Technique to Improve the Search Latency of AND Queries for Large Scale Textual Documents", *2020 IEEE/ACM International Conference on Big Data Computing, Applications and Technologies (BDCAT)*, Leicester, UK, 2020, pp. 37–46.

J. Nabi, "Machine Learning — Text Processing", *Towards Data Science*, n.d., https://towardsdatascience.com/machine-learning-text-processing-1d5a2d638958. [accessed: 03-Mar-2021].

R, "The R Project for Statistical Computing", n.d., https://www.r-project.org/. [accessed: 31-Mar-2021].

D. Sayce, "The Number of Tweets per Day in 2020", n.d., https://www.dsayce.com/social-media/tweets-day/ [accessed 2020].

Scala, "The Scala Programming Language", n.d., https://www.scala-lang.org/. [accessed: 31-Mar-2021].

J. Su, T. Hong, J. Li, J. Su, "Personalized Content-Based Music Retrieval by User-Filtering and Query-Refinement", *Proc. Conference on Technologies and Applications of Artificial Intelligence (TAAI)*, 2018, pp. 177–180.

H. Wu, G. Li, L. Zhou. "Ginix: Generalized Inverted Index for Keyword Search", *Tsinghua Science and Technology*, Vol. 18, Issue 1, 2013, pp. 77–87.

J. Zhang, P. Pu, "Refining Preference-Based Search Results Through Bayesian Filtering," *12th International Conference on Intelligent user Interfaces (IUI), Association for Computing Machinery*, 2007, pp. 294–297.

C. Zhang, H. Wang, Y. Liu, D. Wu, Y. Lia, B. Wang "Automatic Keyword Extraction from Documents Using Conditional Random Fields," *Journal of Computational Information Systems*, Vol. 4, Issue 3, 2008, pp. 1169–1180.

10

Sensor-Based Systems

10.1 Introduction

The ready availability of low-cost sensors and the rapid growth in Internet of Things (IoT) technology is fueling the use of sensor-based systems, including cyber-physical systems. Examples include senor-embedded bridges, sensor-equipped smart buildings, smart machinery, and remote patient monitoring systems. Although sensors may include limited processing capability, the sensor-based smart system often leverages the computing resources in a cloud for processing sensor data as well as its storage resources for data storage. Such processing may be required for monitoring the state of the system as well as for controlling its operations. The processing may be performed both by an edge cloud which is located close to the sensors or on a remote backend cloud or both.

Sensor monitoring and intelligent management of these facilities are targeted at reducing the maintenance cost as well as enabling safe operation of the facilities to prevent failures leading to accidents that may result from the inability to detect faults in a timely manner.

Irrespective of the exact nature of the system, the management of a smart facility typically involves the three following steps:

i. Sensing of the respective phenomena of interest and collecting the sensor data reflecting the current state/health of the facility.
ii. Analyzing the collected sensor data.
iii. Making decisions regarding the management/maintenance of the facility.

A general architecture for cloud-based smart facilities management is presented in Section 10.1.1. A discussion of middleware services required for such management is the focus of attention in Section 10.2. Various sensor-based smart facilities are described in this chapter. Section 10.3 focuses on the management of sensor-equipped bridges, while Section 10.4 discusses a collaboration platform for the research and development of sensor-based smart facilities. Wireless sensor

Resource Management on Distributed Systems: Principles and Techniques,
First Edition. Shikharesh Majumdar.

networks are often an important component of a smart facility that also uses a computing platform such as a grid or cloud for processing the sensor data acquired from the sensor network. Resource management on wireless sensor networks is discussed in Section 10.5. Both allocation and scheduling are important in the context of the management of wireless sensor networks. Scheduling is discussed in Section 10.6, whereas Section 10.7 focuses on sensor allocation. Both algorithms and results of the performance analysis of these algorithms are presented in these two sections. Section 10.8 summarizes the highlights of the discussion presented in this chapter.

10.1.1 Architecture for Cloud-Based Smart Facilities Management

An architecture for smart facilities management discussed in Majumdar (2016) is described next (see Figure 10.1). The figure corresponds to a sensor-embedded bridge, but the architecture can be easily adapted to other types of systems using a diverse set of resources that may include computing and storage servers, software tools for data analysis, archival storage systems, and databases holding various information about the facility including its maintenance history and data repositories. The system administrator, bridge engineer, and bridge operator in Figure 10.1 are personnel who are involved in overseeing the management of the smart facility. Various types of access networks, the backbone network, and the Internet provide intercommunication among various system components and users. A system for

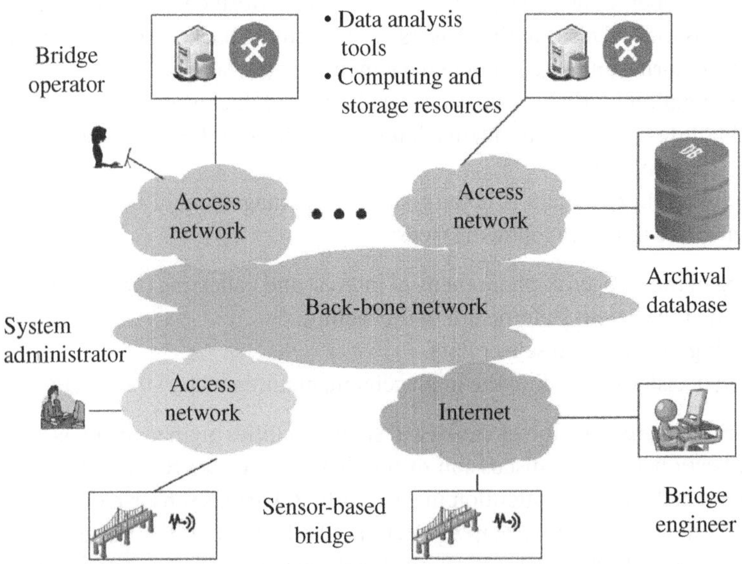

Figure 10.1 Sensor-embedded bridge management. Source: Majumdar (2016)/with permission of Elsevier.

the monitoring of smart bridges for its structural health assessment is described in Lau et al. (2013). Such systems can be useful for analyzing data from sensors on bridges to determining the appropriate time for maintenance of the infrastructure.

Mobile edge computing is often utilized in remote health applications. A smart facility for remote monitoring of chronic patients with the help of both edge devices and a backend server is described in Majumdar (2016). The patient is typically fitted with biological sensors and the data is initially analyzed on a smartphone (edge device) carried by the patient and the identification of critical events, when identified by the edge device, are sent to the backend hospital server for further processing and sending appropriate notifications to medical staff. A more detailed discussion of such a system is presented in Chapter 8.

10.2 Middleware Services

A generic layered architecture for a middleware for use in a smart facility is presented in Figure 10.2 (Majumdar 2016). The system comprises a network, security, and messaging layer. The network layer is used by the messaging layer for intercommunication among the various system resources that include both hardware resources such as computing and storage devices as well as data analytics software.

The security layer used by the messaging layer ensures that communication is performed in a secure manner. A broad range of support for security may be provided: from using virtual private networks to data encryption. The middleware layer and the interface set enable various services for the smart facility and are discussed next.

Authentication and Authorization Service: Authentication is critically important to protect the system from unauthorized users and hackers. Every valid user may not be able to perform all operations and the system needs to verify whether a given user is allowed to perform an operation before the operation is permitted. Addressing the challenges of authentication and authorization on IoT-based systems is well discussed in the literature. Examples of such challenges are highlighted through an example of a hacker unlawfully accessing an insulin pump by impersonating the authorized user (Madsen 2015). Addressing the authentication and authorization issues in the context of sensor-based IoT systems is an important topic for both current and past research. Examples include techniques for the efficient key establishment in cryptography and a role-based Access Control Policy (Liu et al. 2012). Devices such as smart cards and physically unclonable functions for authentication are discussed in Crossman and Liu (2015) that proposes workflows of authentication protocols and describes their behavior on an IoT testbed.

Data Transfer Service: This service is used for performing the transfer of data between system components: a source and a sink. The system needs to support

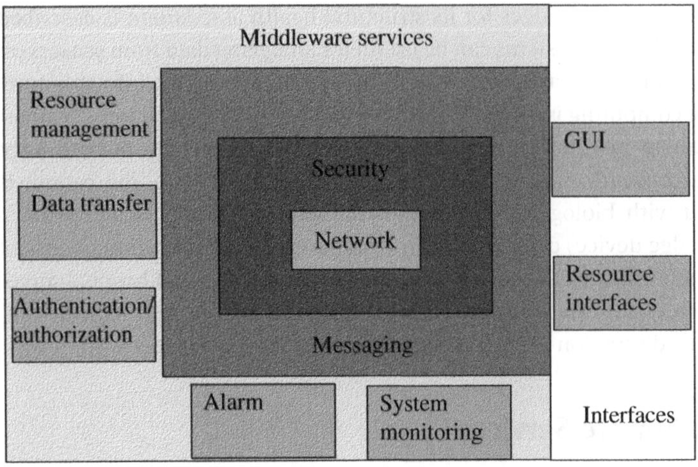

Figure 10.2 Layered architecture of middleware. Source: Majumdar (2016)/with permission of Elsevier.

multiple types of data transfers. Two types of data transfer operations: bulk data transfers and real-time data transfers, for example, are considered in McGregor et al. (2015). During bulk data transfer, a large volume of stored data may be transferred between an originating and a receiving system component. Bulk data transfers are generally performed on a best-effort basis: minimization of transfer latency is important, but there are no deadlines associated with such a data transfer. A deadline is often associated with a real-time data-transfer operation. Moreover, on streaming systems that require real-time data analytics, continuous streams of data need to be analyzed in real time. Resource reservation in advance may be required to meet deadlines on such systems.

System Monitoring Service: The objective of this service is to monitor the health of the different resources (both hardware servers and software tools). Periodic probing is often used to determine whether or not the respective resources are in running condition. Upon discovery of a failure, the system administrator is informed so that corrective action may be performed. System monitoring tools also maintain various performance statistics, such as the utilization of various resources, message transfer rates, and system throughput at various points in time. The measured metrics are then displayed on a dashboard, for example, on request to the system administrator.

Research is directed at addressing the challenges in providing a service for monitoring IoT-based systems that are rooted in the limited power and capabilities of providing flexible communication interfaces and coverage for wireless sensors (Huang and Hsieh 2013). Research is underway to address these challenges. Examples include an ARM9 processor-based gateway for collecting continuous, periodic, and sequential spatial telematics (Huang and Hsieh 2013) and techniques for remote monitoring of an electric vehicle (Jianguo et al. 2013). The frequent emergence of new papers in the area demonstrates the importance of the system monitoring service in the management of sensor-based systems such as smart facilities. A detailed survey, however, is beyond the scope of this book chapter. The interested reader is referred to the list of related papers included in the two papers referred to in this paragraph.

Alarm Service: This service generates an alarm when the analysis of sensor data indicates a fault or an exceptional situation that needs further attention. Sensor data is analyzed and an alarm condition is said to have occurred when a certain predefined system state is attained. The service is responsible for notifying authorized personnel (e.g. the system administrator and/or a facility operator) when such an alarm condition is reached.

Resource Management and Discovery Service: Sensor-based systems use sensor resources provided by a sensor network for example. Resource management in sensor networks is discussed in Section 10.5. Often sensor-based facilities use additional resources such as computing and storage resources (file servers and sensor data repositories) as well as software resources that include various tools for manipulating sensor data and their analysis.

Resources may be localized or distributed. A cloud is often used to provide the glue and unify geographically dispersed resources for a set of authenticated users. Such a cloud-based smart facility for the structural health assessment of bridges is described in Majumdar et al. (2012) and for resource and data sharing for a consortium of mechanical and aerospace engineers in Majumdar (2016). Resources may be acquired on demand as well as reserved in advance. For on-demand (OD) requests, the resource management middleware satisfies the request if the resource is available. The request however is served on a best-effort basis. The resource manager attempts to provide a low latency for request service, but no upper bound on latency is guaranteed. Resources may be requested by users one at a time or multiple resources may be held by a user or application simultaneously. Advance reservation of these resources is required to ensure that all the requested resources are available at the same time. For advance reservation (AR) requests the user is allocated the required resource (if available) for a fixed duration of time between a start and end time specified by the user. When multiple resources need to be booked simultaneously, advance reservation of these resources need to be made.

For smart systems that require real-time responses, a deadline is associated with the AR. Such an AR is characterized by an earliest start time, an estimated execution time for the request and a deadline specified by the user. The resource manager needs to allocate and schedule the request in such a way that the request is completed by the deadline. If this is not possible to achieve in the current state of the system the request is rejected. Resource management for systems subjected to both AR and OD requests is discussed in Chapter 3.

Resource discovery is an important component of resource management. The user provides the system with specific resource attributes and the system generates the list of resources that match these desired attributes. Both the location and the method of accessing them are sent back to the user that originated the resource discovery request.

The interfaces that are typically required for the operation of the smart facility management system are described next.

Graphical User Interface (GUI): The GUI makes the various functionalities for managing smart facilities available to the user/system administrator. Through the various buttons on the GUI, one can perform the desired operations on the system.

Resource Interfaces: A resource such as computing and storage servers as well as software tools are integrated with the platform for facilities management through interfaces, also known as adapters (Majumdar et al. 2012). An adapter provides a common application programming interface that can be used by programs that may be implemented using diverse technologies. For example, a client application running on top of a Linux operating system can invoke a software application running on the Windows platform. Majumdar et al. (2012) and Kanagasundaram et al. (2012) discuss the exposing of resources connected to the cloud-based platform as Web services (WSs). Any operation to be performed on the resource is achieved through the respective WS. Both a SOAP-based as well as a REST (Representation State Transfer)-ful WS can be used for the purpose. A hybrid WS that chooses between these two WS types based on the operation to be performed is discussed in Kanagasundaram et al. (2012). The lightweight feature of a RESTful WS is combined with the support for security and atomicity that accompanies a SOAP-based WS in such a hybrid WS. The research presented in Kanagasundaram et al. (2012) demonstrates that a significant improvement in performance over a standard SOAP-based WS can be achieved by using a hybrid WS.

10.3 Sensor-Based Bridge Management

Bridges are vital infrastructures for a city. The assessment of the state of bridges is important for their maintenance: a fix in time can reduce the cost of repair and prevent bridge failures that are not only expensive to repair but they can also cause loss of human lives. Sensors built into these infrastructures can be monitored to assess the structural health of the bridge. Several billions of dollars are spent annually in Canada and the United States for bridge maintenance. Using a smart facility for the purpose can significantly reduce this expenditure. The architecture of a cloud-based system for building such a smart facility is described in Majumdar

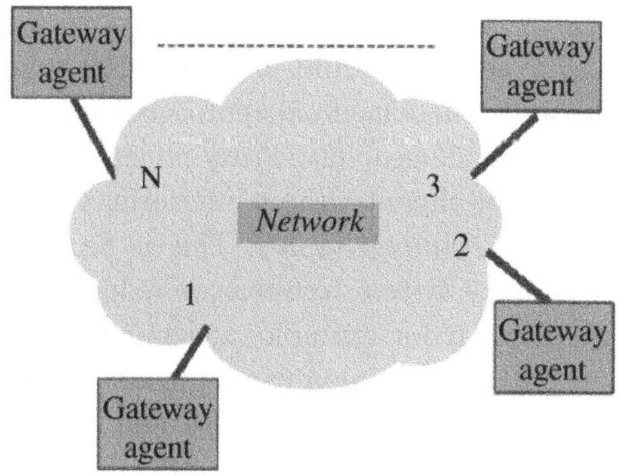

Figure 10.3 Cloud-based smart facility management. Source: Majumdar et al. (2012)/Association for Computing Machinery.

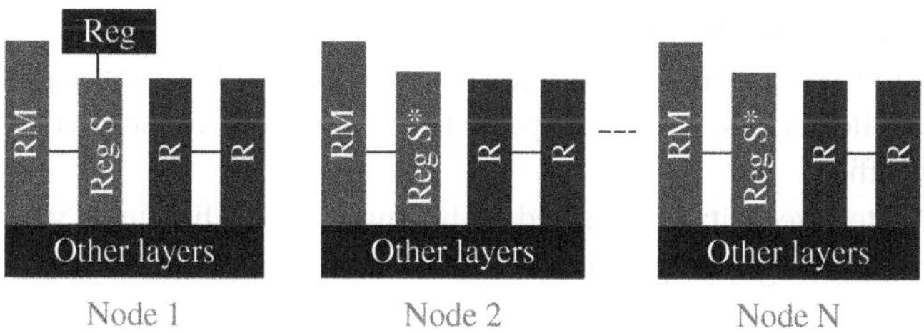

Figure 10.4 Example GWA architecture. Source: Majumdar et al. (2012)/Association for Computing Machinery.

et al. (2012). Resources used by the facility are geographically distributed and are connected by a network (see Figure 10.3). A cloud provides the glue for accessing these resources through gateway agents (GWAs) each of which corresponds to a node in the cloud providing a set of resources.

GWAs have a pluggable architecture for running multiple services (see Figure 10.4).

Each node is equipped with a GWA that deploys two types of services: management services and resource services. Different GWAs can incorporate different types of management and resource services. Reg S is a single resource registry service running on Node 1 and provides information about the location of the various resources on the system. A proxy for the registration service Reg S* is available at each node so that the users at the respective node can access the registry service running on Node 1. Each Service R at a GWA is used for accessing a specific resource available at the GWA. RM is the resource management service that manages the resources available at the given node. Heterogeneity in the implementation of modules is handled by using adapters for interfacing the respective services with the middleware platform.

A number of key features of the system are summarized next. More details are available in Majumdar et al. (2012).

Flexibility: is provided through proxy-pluggable modules that allow the role of an RM-service module, for example, to change with time. This is useful in altering a resource management approach from centralized to distributed and vice versa.

Module Pluggability: Proxy-pluggable modules allow for service implementations to be altered without interfering with the code of other independent services.

Check Pointing Facility: This is provided by a Check Pointing Service. The wastage of system resources is reduced by using checkpoints during large data transfers. If for example, an emergency response interrupts the data transfer, the data transmission can be resumed using the latest check point, thus avoiding the waste of CPU cycles and network bandwidth consumed during the transfer of the previous part of the data.

RESTful Web Services: These are used in performance-critical use cases that correspond to an emergency response, for example.

Location of Domain Experts: This is performed with the help of WSs running on GPS-based smart mobile devices. Thus, an expert can be contacted even when they are away from their office.

Authentication: System security is achieved by having users authenticate with the system before performing any operation.

10.4 Research Collaboration Platform for Management of Sensor-Based Smart Facilities

Research collaboration often involves researchers at multiple locations and research collaboration on sensor-based systems is no exception. The researchers need to share various resources that may be located at different sites each of which corresponds to a subset of researchers in the research team. Such resources include computing servers for running applications processing sensor data, sensor data repositories that are often placed close to the smart facility, archival databases containing the maintenance history of the smart facility, and software tools and applications. These components need to be integrated so that they can be made available to all the members of the collaboration team, irrespective of their locations on demand.

Research platform for smart facilities management (RP-SMARF) described in this chapter can support collaboration among multiple participants by unifying geographically scattered resources from various sites. In addition to supporting a single research consortium RP-SMARF can host multiple collaboration teams at the same time.

A SMARF Control Sever (SCS) is the key element used for controlling the RP-SMARF platform. Resources that belong to the different team members from different sites run behind a firewall provided by the respective institution.

An SMARF Remote Agent (SRA) runs behind every firewall and is responsible for the communication between SCS and the resource. The key features of RP-SMARF are summarized.

- *Multi-Tenancy:* RP-SMARF supports the coexistence of multiple research collaboration teams at the same time. An example use case described in McGregor et al. (2015) provides support for two communities: the bridge management research community (Civil Engineers) and industrial and aerospace machinery research community (Mechanical and Aerospace Engineers). Each community is characterized by its own set of resources that can be used by only the members of the respective community. RP-SMARF provides the required isolation. Although both communities are hosted at the same time on the platform, the resources of a given community are not visible to the members of the other community.

Object model filtering is used by SCS to achieve isolation in the multi-tenant RP-SMARF. The relationship between a user and the respective research team is represented in the object model through the user's settings. When a request for viewing available resources (e.g. computing and storage servers, data repositories, and tools) arrives on the system, the returned list is a subset of all the resources that belong to the user's community. The object model supports the tagging of resources using "tag" objects, which support the tagging of resources with additional metadata. The primary use of the tags is to specify the community that owns a resource or tool. The metadata for a resource that contains the access rights for various users on the system is stored in another object. This object is used for deciding whether or not a requesting user has the right to perform the desired operation on the resource.

- *Support for Software Tools:* Collaborators often devise software and specialized tools for supporting the research performed in the consortium. RP-SMARF provides a graphical user interface (GUI) that enables the easy integration and use of such tools and application software with the platform. It can be used to set up a tool facility that also provides a walkthrough for a user through the steps of configuring and running the respective tool on any of the sites integrated by RP-SMARF. This enables a quick learning facility for the user who can avoid the lengthy period of time they will need otherwise to learn to configure and use the tool.
- *Heterogeneity and Multiple Modes of Operation:* The RP-SMARF platform also supports the incorporation of various software tools, the implementation of which may be based on different technologies. Some tools (McGregor et al. 2015) allow dual modes of operation. In the interactive mode the user can interact with the tool and select the operation to be performed on a given set of

data set. In the batch mode, the tool can be used to process a batch of multiple data files. SRA is used for intercommunication between the user and a tool.

Interactive tool invocation is achieved by using a Virtual Network Computing (VNC) server or the Windows Remote Desk Top Protocol (RDP) running on the remote system deploying the tool. For a user to connect to the machine hosting the tool, the user's browser connects to a Guacamole server on SCS (discussed in more detail in the following paragraph) that passes the commands for performing the desired operations on the tool to the respective SRA. Only one user can interact with the tool at a time in this mode. If the tool is already being used the user request is rejected.

There is no Guacamole session and the user cannot directly have a remote session on the computer executing batch commands. The input data, the output folder where the results are to be collected and the tool itself may be physically located in separate locations. The communication between the user's browser and SCS is based on REST. When SRA invokes the tool, it copies the input data to the machine running the tool and the resulting data generated by the tool is copied to the output folder identified by the user. Similar to the interactive tools, the platform supports multiple batch tools, hence allowing heterogeneity.

- *System Flexibility Allowing Support for Dynamic Attaching and Detaching of Tools:* A tool owner can change the availability of a tool dynamically. An "Attach" or "Detach" operation invoked through the GUI results in adding access to the respective tool or removing such access dynamically during system operation.
- *Support for Data Discovery Through Metadata Searching:* Metadata describes the data, sensor data, for example, to be analyzed. With hundreds of such data files containing sensor data collected on a smart facility, an effective search of the datasets requires using domain-specific fields to search data. For example, wind speed and wind direction may be used to search for vibration data files for a bridge (Majumdar 2016) that correspond to specific days and times. All the files containing vibration data that match the specified metadata ranges provided by a researcher will be identified at the end of the search, thus avoiding a laborious manual search by the researcher through hundreds or thousands of data files collected on the bridge. RP-SMARF users define metadata by using software based on the Semantic Web Resource Description Framework (RDF) (W3C – 2014).
- *Data Transfer Modes:* Two modes of data transfer are supported. In the *bulk data transfer mode* multiple files are transferred from the source to the destination that are selected by the user. SCS receives the bulk data transfer request via REST and forwards it to the source SRA that has the data. A command is then issued by the source SRA to transfer the data from the source data repository to the

destination folder. At the end of this data transfer it contacts the destination SRA that transfers the data to the desired (destination) data resource.

Streaming data transfer mode is used in use cases such as the one in which data generated by an industrial machine in a lab needs to be analyzed by a tool that is located on a different site. This data transfer service enables the transfer of data in real time from a local or remote site to the server running the tool used for analyzing the data.

Additional features of RP-SMARF that include tool/data set discovery and resource reservation are described in McGregor et al. (2015).

Key Components of RP-SMARF are presented in Figure 10.5. As indicated earlier, it integrates resources running on multiple nodes (e.g. Institution X and Institution Y) of the cloud and is controlled by a single control server (SCS) that uses the Interactive Connectivity Establishment (ICE) technique to talk to other system components. Each node comprises various resources (e.g. domain-specific computers and domain-specific file servers). In addition to running various middleware services, SCS contains a database of metadata including resource characteristics and their access rights. The control server (SCS) in the public cloud in Figure 10.5 can be scaled horizontally when required to improve availability and handling an increase in workload intensity. SCS stores all information related to system control in an SQL database. The client providing a GUI through a browser is implemented in JavaScript using the AngularJS framework. The browser uses the Hyper Text Transfer Protocol (HTTP) for communication. The *remote agent* is able to use the secure Shell (SSH) protocol/Secure File Transfer Protocol (SFTP) for communicating with the node which runs the application (or holds the data). It has a file system interface that facilitates user interaction with the system and a job-running service used by the agent to start and stop jobs. The *Guacamole server* (Apache Guacamole n.d.) is a web server that facilitates interactive GUI-based access to resources through the remote desktop protocol (RDP) for Windows-based systems and via virtual network connection (VNC) for other systems. Further details of the system are provided in McGregor et al. (2015).

10.5 Resource Management on Wireless Sensor Networks

Smart facilities discussed earlier in the chapter are often based on sensor networks. Due to their ready availability and cost-effectiveness, wireless sensor networks are often deployed on such smart facilities. Management of wireless sensor networks accessed by a single client is a well-studied problem. With the advancement of sensor technology and with the aim of increasing the cost

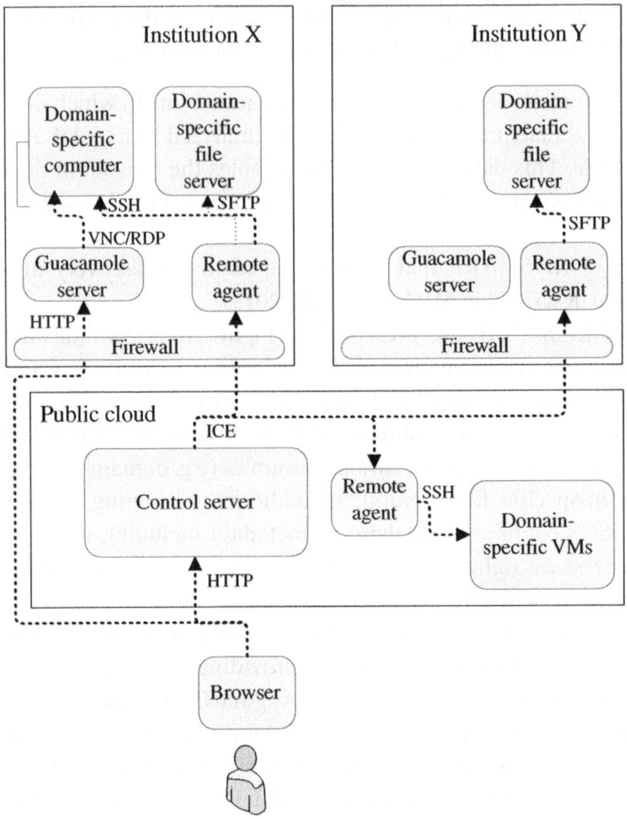

Figure 10.5 Key components of RP-SMARF. Source: McGregor et al. (2015)/with permission of IEEE.

efficiency of wireless sensor networks, multipurpose wireless sensor networks have been replacing sensor networks dedicated to a single application (Akbar et al. 2006; del Cid et al. 2010). However, sensors in a system may need to be shared among multiple applications. Such systems often deploy multi-function sensor nodes for sensing multiple phenomena. Multiple client applications running on a smart facility management system, each using a specific type of sensor data, can access the sensor nodes that are shared among these various applications at the same time. The resource management problem in this case is similar to the resource management problems discussed earlier in Chapter 3. The algorithms for resource management that need to deal with the allocation of sensors to competing applications and the scheduling of sensor requests that wait in a queue for the same sensor node are important. Both allocation and scheduling techniques in the case of a sensor network attached to a grid or cloud

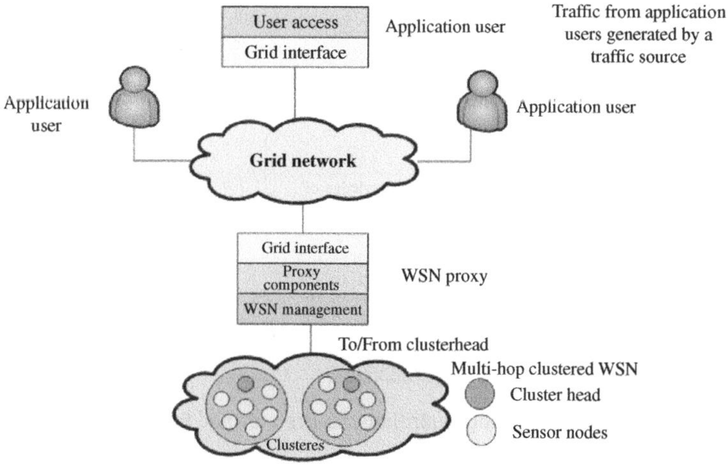

Figure 10.6 WSN and the grid. Source: Kapoor et al. (2012b)/with permission of IEEE.

have been described in (Kapoor et al. 2012b, 2015) are discussed. The model of a system that combines a Wireless Sensor Network (WSN) with a grid equipped with computing and storage resources for processing the sensor data is presented in Figure 10.6 (Kapoor et al. 2012b). Sensor grids combine Wireless sensor networks (WSNs) with a grid to integrate the detection of the specific phenomena of interest with the processing of the detected sensor data performed on the grid (Chandrasekaran et al. 2014; Tham and Buyya 2005). Although Figure 10.6 refers to a grid the same functionality can be achieved with the help of a cloud as well. In line with grids, cloud infrastructures can also provide computing, storage, and processing for data obtained from wireless sensor networks (Amri et al. 2013).

Users send requests for getting the desired information from the WSN via the Grid Network (see Figure 10.6). The proxy in the figure performs two key operations: it exposes the functionality of the sensor nodes to the grid and runs the resource management algorithms (WSN management). The resource management functionality of the proxy node (see Figure 10.6) can correspond to the resource management service shown in Figure 10.2, for example. Computing and storage resources in the grid are used for running data analytics applications for the smart facility management system. A WSN is often deployed in the form of clusters of sensor nodes. The communication between the proxy and the other sensor nodes in the cluster is handled by a specific node called the cluster head (see Figure 10.6). Messages from the proxy reach the cluster head first and then travel to the designated receiving sensor node via multiple hops through the sensor nodes that lie in the path. A discussion of scheduling of user requests queued up for a given sensor is discussed in Section 10.6 and sensor allocation is discussed in Section 10.7.

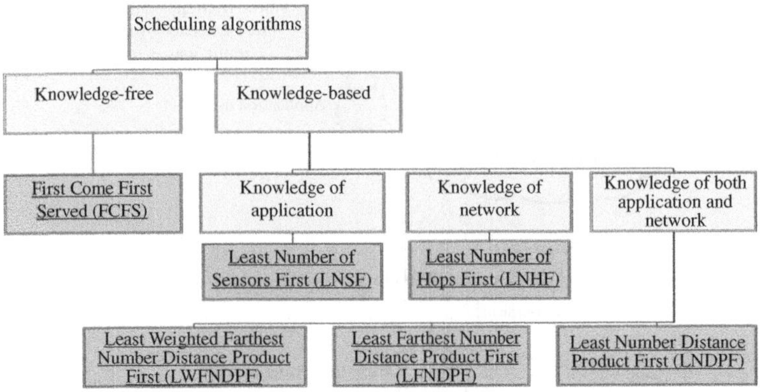

Figure 10.7 Description of scheduling algorithms. Source: Adapted from Kapoor (2013).

10.6 Scheduling on WSNs

As discussed in Chapter 3, scheduling refers to the determination of the order in which requests are to be processed when multiple requests are requesting the resource. On WSNs, at a given point in time, multiple requests can compete with one another for using a specific sensor. The job of the scheduling algorithm running at the proxy is to decide the order in which these requests will run on the given sensor. The scheduling algorithms discussed in this chapter are divided into two categories: one that does not use any information about request characteristics and one that is based on sensor characteristics (see Figure 10.7). These are described next.

10.6.1 Knowledge Free Algorithms

First-Come-Frist-Served (FCFS) is a knowledge-free algorithm that does not use any knowledge of senor or application characteristics. Under this scheduling algorithm, requests are scheduled in an order that is the same as the order in which they arrived.

10.6.2 Knowledge-Based Algorithms

These algorithms use one or more characteristics of the application generating the request and/or the sensor network. Examples of such characteristics are discussed first in the next subsection and the descriptions of the algorithms that use these characteristics are discussed in the following subsection.

Application and WSN Characteristics:

Number of Sensors requested (N): is the total number of sensors required by the application.

Distance: is the number of hops required by a message corresponding to a request to travel from the cluster head to a given sensor node.

10.6.2.1 Description of the Knowledge-Based Scheduling Algorithms

Three types of scheduling algorithms are described: those that are based only on the knowledge of the application (request), those that are based on only the knowledge of the network, and those that use combined knowledge of both the request and the network. Each algorithm corresponds to a specific type/types of knowledge of the application request/network it uses. The algorithms for these three categories are presented next.

Knowledge of Application-Based Algorithm

Least Number of Sensors First (LNSF): When multiple applications compete for the WSN, this algorithm selects the application that is characterized by the least number of sensors to run first.

Knowledge of Network-Based Algorithm

Least Number of Hops First (LNHF): The sensors requested by an application can be at different distances from the cluster head. With this algorithm the application request that is characterized by the least average distance of the sensors from the sensor head is selected to run first.

Combined Knowledge of Application and Network-Based Algorithms

The algorithms that use both the application and the network are described next.

Least Number Distance Product First (LNDPF): Applications vary both in terms of the number of sensors required (*N*) and the average distance of the sensors allocated to the application. *Number Distance Product (NDP)* of an application is defined as the product of the number of sensors required and the average distance of the sensors from the cluster head. With this algorithm, requests are scheduled in nondecreasing order of NDP for the respective applications.

Least Farthest Number Distance Product First (LFNDPF): The messages to and from the farthest sensors result in maximum delays in the WSN. *Farthest Number Distance Product (FNDP)* is the product of the number of sensors that are located at the farthest distance (in terms of hops) and this farthest distance. With this scheduling algorithm, the focus is only on the sensors that are located farthest from the cluster head and requests corresponding to the respective applications are scheduled in nondecreasing order of FNDP of the applications.

Least Weighted Farthest Number Distance Product First (LWFNDPF): This algorithm differentiates between sensors based on their distance and associates a weight with each hop distance. *WFNDP* of each application is obtained by multiplying the weight for a particular hop distance with FNDP defined in the case of the previous algorithm. The requests from applications are scheduled in nondecreasing order of their WFNDP.

The performances of the scheduling policies are discussed in the following subsection.

10.6.3 Performance of Scheduling Algorithms

A simulation-based performance analysis of the algorithms reported in (Kapoor et al. 2011, 2012b) is presented. The open system is subjected to request arrivals for applications that follow a Poisson process. The various characteristics of the workload and system are presented in detail in the respective papers. Only a discussion of the performance results is presented in this section. A factor-at-a-time experimental procedure is followed: any single factor is varied in an experiment while the others are held at their default values. The details of the simulation models and workload and system parameters are available in (Kapoor et al. 2011) for experiments that focus on scheduling algorithms based on a single characteristic: of the application or of the network. Algorithms that use the knowledge of both application and system characteristics are presented in Kapoor et al. (2012b).

10.6.3.1 Performance of Algorithms That Use a Single Characteristic

The performance of LNSF that uses the knowledge of the application characteristic, *N*, the number of sensors used by the application, and LNHF that uses knowledge of the average distance (measured in terms of the number of hops) between sensors and the cluster head are compared with the performance of the knowledge free algorithm FCFS. Both these characteristics are found to be useful in scheduling. Figure 10.8 shows the superiority of LNSF over FCFS for all the arrival

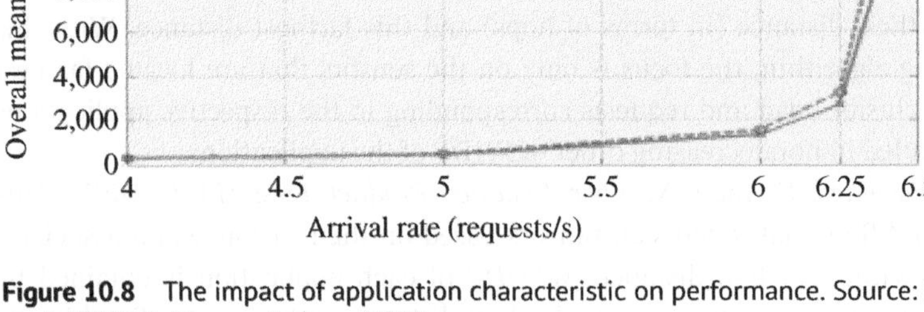

Figure 10.8 The impact of application characteristic on performance. Source: Kapoor et al. (2011)/with permission of IEEE.

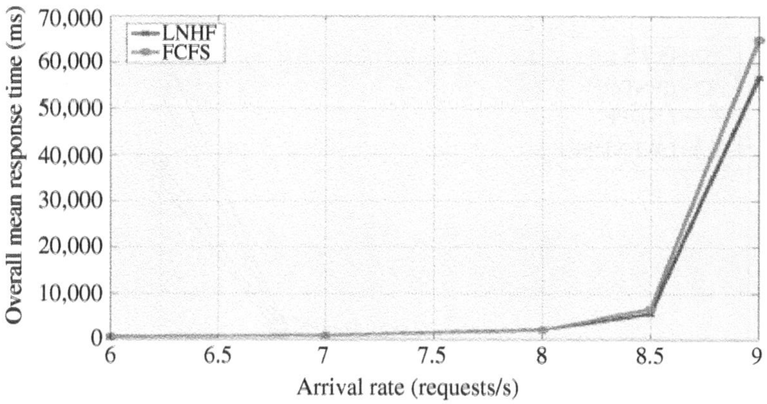

Figure 10.9 The impact of system characteristic on performance. Source: Kapoor et al. (2011)/with permission of IEEE.

rates experimented with. The advantage of using the average distance of sensors is demonstrated in Figure 10.9 in which LNHF exhibits a superior performance in comparison to that of FCFS that does not use the knowledge of any application or system characteristics. As the arrival rate increases, the resource contention increases and the system gets stressed more. The advantage of using knowledge of the application/system characteristics is observed to increase with an increase in system contention (see Figures 10.8 and 10.9). As explained in (Kapoor et al. 2011), further investigation revealed the advantage of using more than a single parameter in scheduling. This is discussed in the following section.

10.6.3.2 Performance of Algorithms That Use the Knowledge of Both Application and System Characteristics

The performance of LNDPF, LFNDPF, and LWFNDFP that use both application (request) and system (WSN) characteristics is compared with that of the knowledge-free algorithm FCFS.

Effect of Arrival Rate (λ) The arrivals were equally divided among the applications used in the experiments. Thus, with A applications, each application generated an arrival rate of λ/A. The overall mean response time is the average of the response time for all the requests that arrive in the system. Recall that the response time of a request is the difference between the time the request processing is completed and the arrival time of the request. The impact of λ on performance is captured in Figure 10.10. As the arrival rate increases, contention for system resources increases leading to an increase in mean response time for each algorithm.

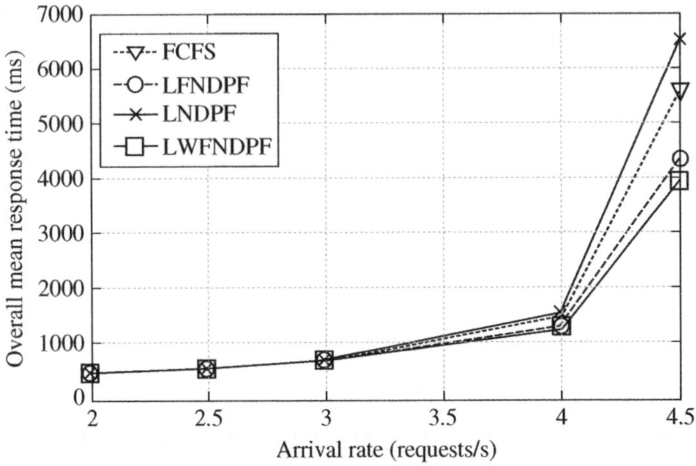

Figure 10.10 Effect of arrival rate. Source: From Kapoor et al. (2012b).

LNDPF performs the worst, while LWFNDPF performs the best, followed by LFNDFP. The inferior performance of LNDPF in comparison to FCFS demonstrates the requirement for the use of appropriate system and workload characteristics in scheduling. The simulation results indicate that for the system and workload parameters experimented with, the number of sensors at the farthest distance from the cluster head plays a more important role in comparison to N, the total number of sensors requested by the application. This is demonstrated by the superior performance achieved by LFNDPF and LWFNDPF.

Effect of Data Size Intercommunication delays are known to be a significant component of the request processing times in sensor networks. Larger data transfers are expected to give rise to larger delays. The impacts of the size of the request and response messages are captured in Figure 10.11. The sizes of all request and response messages are held at the same value defined by the data size parameter that is varied from 50 to 150 bytes. The workload intensity increases with an increase in this workload parameter, increasing system contention and consequently increasing the mean overall response time. Once again, LWNDPF performs the best.

The Effect of Number of Applications So far, LWNDPF performs the best in all the experiments discussed earlier. Whether this conclusion holds for various numbers of applications is investigated in another experiment. The results of this experiment confirm that LWNDPF leads to a superior performance in comparison to the three other scheduling policies for various numbers of applications that share the WSN (see Figure 10.12).

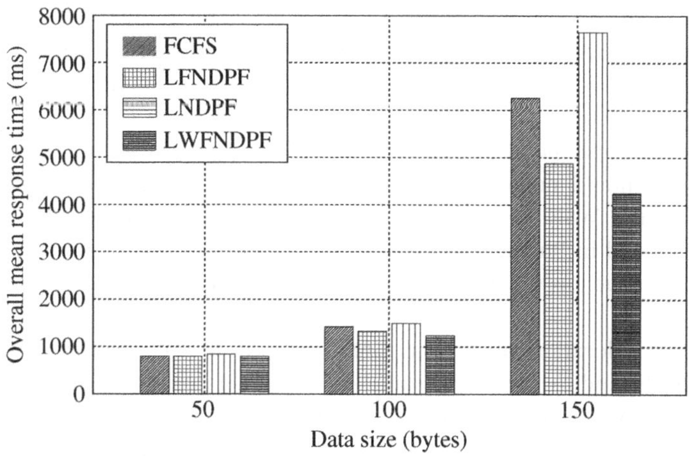

Figure 10.11 Effect of data size. Source: Kapoor et al. (2012b)/with permission of IEEE.

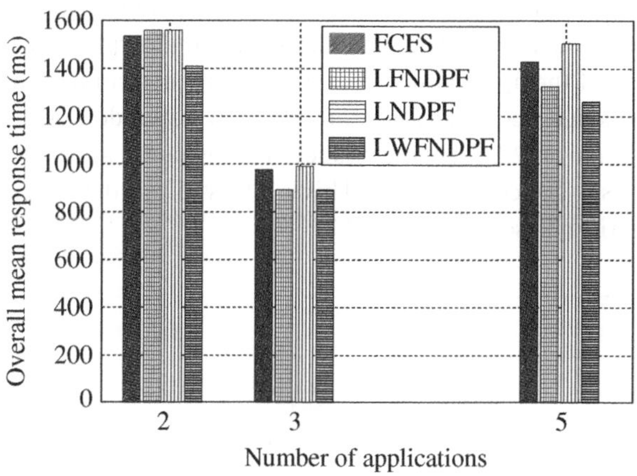

Figure 10.12 Performance results with various numbers of applications. Source: Kapoor et al. (2012b)/with permission of IEEE.

10.7 Sensor Allocation

The allocation of sensors to applications determines which sensors in the WSN are to be allocated to which application. Once the allocation is complete applications send requests for the respective sensors. The scheduler then determines when the application requests are sent to the WSN.

Resource management and sensor allocation in WSNs hosting multiple applications have received attention from researchers. Examples include (Tan and Tong 2007) that focuses on mobile object tracking. The authors advocate the grouping of sensors that can effectively contribute to tracking the target of interest while putting the other sensors in the WSN to sleep. Energy harvesting to prolong the lifetime of a WSN is discussed in Ren-Shiou et al. (2010). The authors of the paper point out that the rate of energy drain from a set of allocated sensors should not exceed the energy harvesting rate to maintain the operation of the WSN. Sharing the sensors in a WSN among multiple applications discussed in Bhattacharya et al. (2010) indicates the importance of systems constraints such as the memory capacity of sensors in sensor allocation algorithms.

Two classes of allocation algorithms are discussed in this section (see Figure 10.13): *static* and *dynamic*. For static allocation algorithms, sensor allocation to applications is performed only once and does not change during the lifetime of the application. For dynamic allocation algorithms, the determination of which sensor among the sensors can serve the request is performed dynamically upon request arrival.

Each of these classes of algorithms can be divided into two subcategories: knowledge-based and knowledge-free. The knowledge-based algorithms use one or more characteristics of the application generating the request and/or the sensor network. The knowledge-free algorithms, on the other hand, do not use any knowledge of application or WSN characteristics. The dynamic allocation algorithms are discussed first (see Section 10.7.1). This is followed by a discussion of the static allocation algorithms (see Section 10.7.2)

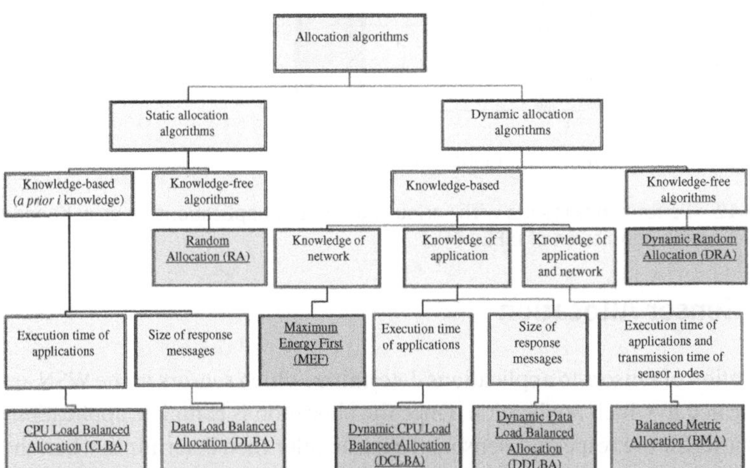

Figure 10.13 Classes of allocation algorithms. Source: Adapted from Kapoor et al. (2012a).

10.7.1 Dynamic Allocation Algorithms

Two classes of algorithms: knowledge-free and knowledge-based algorithms are discussed.

10.7.1.1 Knowledge-Free Algorithm

Dynamic Random Allocation (DRA)

DRA does not use any information about the characteristics of the application or the network. Sensor nodes are allocated at random. Each sensor node in the set of sensor nodes that can be used for serving this application has an equal probability of being selected.

10.7.1.2 Knowledge-Based Algorithms

Allocation Algorithm Based on the Knowledge of Network (WSN)

Maximum Energy First (MEF)

Sensor nodes are resource-constrained. Thus, reducing the energy consumption of these nodes in order to improve the operational lifetime of the wireless sensor network is an important system objective. The MEF algorithm focuses on increasing the network lifetime requires complete knowledge about the energy levels of all sensor nodes. MEF sorts sensor nodes that can be used for satisfying this request in a nonincreasing order of energy available on sensors. From the set of these available nodes, the sensor nodes that have the highest available energy are selected for the execution of the job requests.

Allocation Algorithm Based on the Knowledge of Applications:

Dynamic CPU Load Balanced Allocation (DCLBA)

This algorithm uses information about execution time of application requests. It focuses on balancing the energy consumption due to the CPU component amongst the various sensor nodes. For each sensor node, the proxy maintains information about the sum of execution times of job requests to which the sensor node has already been allocated. The sensor nodes are allocated in nondecreasing order of this sum.

Dynamic Data Load Balanced Allocation (DDLBA)

This algorithm uses information about the sizes of response messages and attempts to balance the energy consumed while transmitting the response messages corresponding to the requests processed by a sensor node. The energy consumed for data transmission is directly proportional to the size of the transmitted message.

For each sensor node, the proxy maintains information about the sum of the sizes of the response messages transmitted by the sensor node. The proxy sorts the list of the sensor nodes in nondecreasing order of this sum. The sensor nodes are allocated in nondecreasing order of the sum.

Allocation Algorithm Based on the Knowledge of Both Applications and the Network

Balanced Metric Allocation (BMA)

This algorithm uses knowledge of energy used both due to request execution and message transmission for requests. The goal of BMS is to balance the energy consumption, both due to the CPU component and the radio component. For the CPU component, the energy consumed due to the execution of requests allocated to a sensor node is considered. This is designated as ECPU. For the radio component, energy consumed while transmitting a message is considered. This is designated as ERadio. With the BMA algorithm, sensor nodes are allocated to job requests in nondecreasing order of the sum of ECPU and ERadio. Determination of ECPU and ERadio is beyond the scope of this chapter. The interested reader is referred to Kapoor et al. (2015) for a more detailed discussion of the topic.

10.7.2 Static Allocation Algorithms

Two classes of algorithms: knowledge-free and knowledge-based algorithms are discussed.

10.7.2.1 Knowledge-Free Allocation Algorithms

Random Allocation (RA)

RA does not use any information about the characteristics of the application or the network. Sensor nodes are allocated at random. Moreover, the sensors allocated to an application are fixed and do not change with time. Each sensor node in the set of sensor nodes that can be used for serving this application has an equal probability of being selected.

10.7.2.2 Allocation Algorithms Based on the Knowledge of Applications

CPU Load Balanced Allocation (CLBA)

This algorithm uses information about the execution time of application requests. It is similar to DCLBA, except that the sensor allocation to applications is static and does not change with time.

Data Load Balanced Allocation (DLBA)

This algorithm uses information about the sizes of response messages and attempts to balance the energy consumed while transmitting the response messages corresponding to the requests processed by a sensor node.

It is similar to DDLBA except that the sensor allocation to applications is static and does not change with time (i.e. from one request to another).

10.7.3 Performance of Sensor Allocation Algorithms

The performance of the dynamic allocation algorithms is discussed first. This followed by a discussion of the performance of the static allocation algorithms.

10.7.3.1 Performance of Dynamic Allocation Algorithms

A simulation-based performance analysis of the allocation algorithms reported in Kapoor et al. (2013, 2015) is discussed in this section. The open system used in these investigations is subjected to request arrivals for applications that follow a Poisson process. The various characteristics of the workload and system are presented in detail in the respective papers. Only a summary of the discussion of the performance results is presented in this book. As in the case of the analysis of the scheduling algorithms discussed earlier, a factor-at-a-time experimental procedure is followed: only a single factor is varied in an experiment while the others are held at their default values. The details of the simulation models and workload and system parameters are available in Kapoor et al. (2013) for experiments that focus on dynamic sensor allocation.

Energy of sensors gets depleted as they continue to get used. A sensor network remains active and usable if all the sensors in the network have enough remnant energy to function effectively. Thus, the aim of the sensor allocation algorithms is to minimize and balance the energy consumption among the sensor nodes in the WSN. The following metrics have been used to evaluate the simulation-based performance of the algorithms presented in Kapoor et al. (2015). The discussion presented explains the impact of the different system and application characteristics on the relative performance of the algorithms that is important for both users and system designers.

For the results presented in this section, the initial energy for all the sensors is assumed to be at the same level at the beginning of the simulation period. Experiments with different initial energy levels for sensors are described in Kapoor et al. (2015).

Minimum Energy (ME): It is the energy of a sensor node that has the least amount of remnant energy at the end of the simulation period.

Network Lifetime (NLT): It is defined as the time when the energy of any sensor node falls below a predefined threshold and the WSN cannot be used.

Effect of Arrival Rate As arrival rate increases, the minimum energy metric for all the algorithms is observed to decrease. More arrivals per unit of time lead to the use of more energy in the sensors that constitute the WSN as a result of which the energy drain at each sensor node increases, leading to a deterioration in ME for the WSN (see Figure 10.14). The knowledge-based algorithms perform superior to DRA, which allocate sensors randomly and does not use any knowledge of application or WSN characteristics. BMA and MEF that use information about both CPU and network characteristics demonstrate a higher performance (higher ME) in comparison to the DCLBA and DDLBA that use the knowledge of only a single characteristic: ECPU or ERadio, respectively.

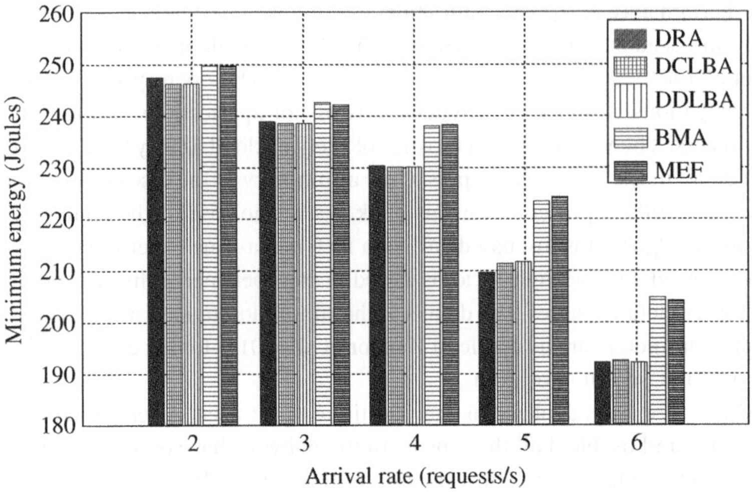

Figure 10.14 The impact of arrival rate on ME. Source: Kapoor et al. (2013)/with permission of IEEE.

Network lifetime is directly related to ME: a lower ME leads to a lower NLT and the relative performance of algorithms observed when using the NLT performance metric (see Figure 10.15) is similar to that obtained when ME is used as the performance metric. Because of the similarity in the conclusions

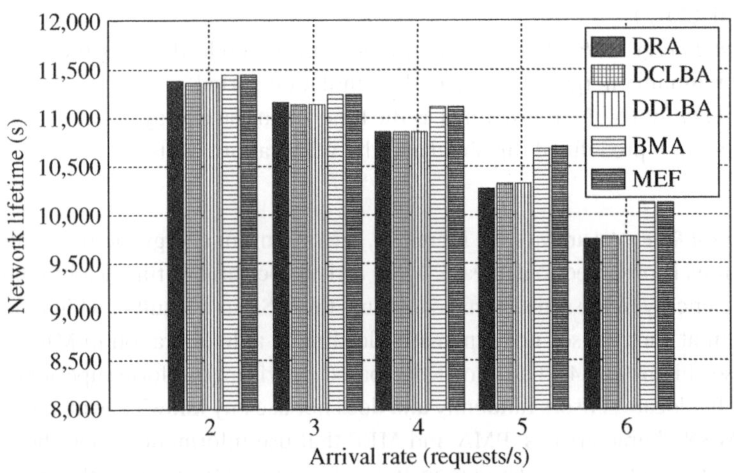

Figure 10.15 The impact of arrival rate on NLT. Source: Kapoor et al. (2013)/with permission of IEEE.

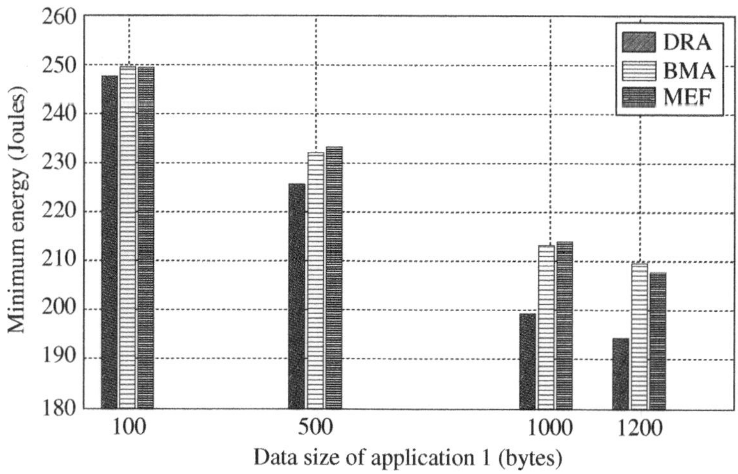

Figure 10.16 The impact of data size on ME. Source: Kapoor et al. (2013)/with permission of IEEE.

derived from Figure 10.14 and Figure 10.15, only the performance results for ME are reported in the following discussions.

Impact of Data Size The performance of two high-performing knowledge-based algorithms, BMA and MEF, is compared with that of the knowledge-free algorithm DRA for different data size (see Figure 10.16). For any allocation algorithm, increasing data size increases the energy used in message transmission. As a result, ME for all algorithms tends to decrease with an increase in data size (see Figure 10.16). The relative performance of the allocation algorithms captured in the figure is similar to that observed in Figure 10.14. A similar impact of data size on the performance of the dynamic allocation algorithms is captured in the figure. BMA and MEF demonstrate a superior performance in comparison to DRA.

10.7.3.2 Performance of Static Allocation Algorithms

As in the case of dynamic allocation algorithms, using knowledge of application and system characteristics in applications is found to be beneficial in the case of static allocation algorithms as well (Kapoor et al. 2012a). The performance of the static allocation algorithms is observed to be inferior to their dynamic allocation counterparts. A comparison of the static algorithm that performs the best (CLBA) and the high-performing dynamic algorithm (BMA) is presented in Figure 10.17. For any given arrival rate, BMA is observed to perform significantly better than CLBA (Kapoor et al. 2015). No further discussion of the performance of static algorithms is thus provided. The interested reader is referred to Kapoor et al. (2015).

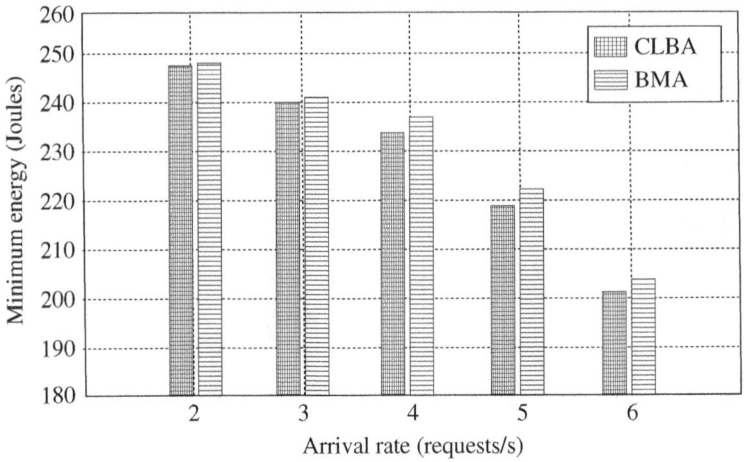

Figure 10.17 Performance comparison of static and dynamic allocation. Source: Kapoor et al. (2015)/with permission of IEEE.

10.8 Summary

Sensor-based systems were the focus of this chapter. Sensor-based smart facilities management was discussed first. Such systems are often used for monitoring the state of the facility and/or controlling it. Examples include sensor-based bridges as well as sensor-equipped smart machinery. The systems often include sensors that monitor various phenomena that indicate the health and state of the smart infrastructure as well as computing clouds or grids that process the sensor data and help with the maintenance and control of the facility. The architecture of a smart facility and the generic middleware services used for its maintenance/control were described. The chapter also includes a detailed discussion of two examples of smart facilities. The first focuses on monitoring and maintenance of bridges that are fitted with sensors that monitor phenomena such as vibrations occurring on the bridge. The second example is a smart facility used for collaboration among members of a given team of researchers that share various data, tools, and servers that are of common interest with one another. A novel feature of this facility is its support for multitenancy which allows multiple research teams, each with its own set of tools and resources to use the platform at the same time.

The later part of the chapter focused on wireless sensor networks that comprise clusters of sensors inter-connected by a wireless network. The senor data is processed by a grid/cloud connected to the WSN though a proxy node. The WSN can be used simultaneously by multiple users, each of whom runs one or more applications that send requests to the sensors in the WSN. Both allocation of sensors to applications and the scheduling of request execution at each sensor node are important resource management functionalities and are discussed in the chapter. Two classes of scheduling and allocation algorithms are described. The algorithms in the first class are knowledge-free in the sense that they do not use any characteristic of the application and the system, whereas the second class of algorithms is based on application and/or system characteristics. A simulation-based performance analysis that provides insights into system behavior and performance is included.

Exercises

1 Discuss the role of clouds in smart facilities management. Do you think both clouds at the edge as well as in the backend can be useful in managing smart facilities? If yes, provide examples in support of your answer.

2 Explain how the RP-SMARF system provides (a) multitenancy and (b) support for heterogeneity.

3 Describe the roles of a cluster head and a proxy node in a wireless sensor network. Discuss how sensing, computation, and communication are supported on a WSN.

4 Discuss the role of using knowledge of the system and applications in the context of both request scheduling and sensor allocation on a wireless sensor network. Provide a list of the various different types of system and application characteristics and discuss the importance of using the knowledge of each of them in resource management.

5 (a) Describe the utility of the two performance metrics: overall mean response time and network lifetime in the context of resource management on WSNs.
(b) Compare the impact of arrival rate on the overall mean response time of all the scheduling algorithms described in the chapter.
(c) Compare the impact of arrival rate on the network lifetime achieved by the different allocation algorithms described in the chapter.

6 Discuss the impact of data size on the performance of all the scheduling and allocation algorithms described in the chapter.

References

A.H. Akbar, A. Ali Iqbal, K. Kim, "Binding Multiple Applications on Wireless Sensor Networks", Advances in Grid and Pervasive Computing (GPC), Chung, Y.C., Moreira, J.E., Eds. Lecture Notes in Computer Science, Vol. 3947, 2006, pp. 250–258.

A.L. Amri, W.S. Ansari, M.M. Hassan, M.S. Hossain, A. Alelaiwi, M.A. Hossain, "A. Survey on Sensor-Cloud: Architecture, Applications, and Approaches", *International Journal of Distributed Sensor Networks*, Vol. 2013, 2013, Article ID 917923, 18 pp., doi: 10.1155/2013/917923.

Apachae Guacamole, n.d., https://guacamole.apache.org/. [accessed: March 2024].

S. Bhattacharya, A. Saifullah, C. Lu, G.C. Roman, "Multi-Application Deployment in Shared Sensor Networks Based on Quality of Monitoring", *Proc. of the 16th IEEE Real-Time and Embedded Technology and Applications Symposium (RTAS '10)*, Stockholm, Sweden, April 2010, pp. 259–268.

S. Chandrasekaran, E. Choi, J.H. Abawajy, R. Natarajan, "Sensor Grid Middleware Metamodeling and Analysis", *International Journal of Distributed Sensor Networks*, Vol. 2014, 2014, Article ID 805708, 12 pp.

P.J. del Cid, S. Michiels, W. Joosen, D. Hughes, "Middleware for Resource Sharing in Multi-purpose Wireless Sensor Networks," *Proc. of IEEE International Conference on Networked Embedded Systems for Enterprise Applications (NESEA)*, November 2010, pp. 1–8.

M.A. Crossman, H. Liu, "Study of IoT with Authentication Testbed", *Proc. of the 2015 IEEE International Symposium on Home Land Security (HLS)*, Waltham, USA, April 2015.

J-D. Huang, H.C. Hsieh, "Design of Gateway for Monitoring System in IoT Networks", *Proc. of the 2013 IEEE International Conference on Green Computing and Communications and IEEE Internet of Things and IEEE Cyber, Physical and Social Computing*, Beijing, China, August 2013.

X. Jianguo, X. Gang, Y. Mengmeng, "Monitoring System Design and Implementation Based on the Internet of Things", *Proc. of the 2013 Fourth International Conference on Digital Manufacturing and Automation*, Qingdao, China, June 2013.

R. Kanagasundaram, S. Majumdar, M. Zaman, P. Srivastava, N. Goel, "Exposing Resources as Web services: A Performance Oriented Approach", *Proc. of the 2012 International Symposium on Performance Evaluation of Computer and Telecommunication Systems (SPECTS'12)*, Genoa, Italy, July 2012.

N.K. Kapoor, "Resource Management in Wireless Sensor Networks Hosting Multiple Applications", Ph.D. Thesis, Department of Systems and Computer Engineering Carleton University, Ottawa, Canada, 2013.

N.K. Kapoor, S. Majumdar, B. Nandy, "Scheduling on Wireless Sensor Networks Hosting Multiple Applications", *Proc. IEEE International Conference on*

Communications (ICC 2011) – Adhoc, Sensor and Mesh Networking Symposium, Kyoto, Japan, June 2011.

N.K. Kapoor, S. Majumdar, B. Nandy, "Sensor Allocation to Multiple Applications in Shared Wireless Sensor Networks", Poster, *31st IEEE International Performance Computing and Communications Conference (IPCCC)*, Austin, USA, December 2012a.

N.K. Kapoor, S. Majumdar, B. Nandy, "System and Application Knowledge Based Scheduling of Multiple Applications in a WSN", *Proc. of the IEEE International Conference on Communications (ICC 2012) – Adhoc, Sensor and Mesh Networking Symposium*, Ottawa, Canada, June 2012b.

N.K. Kapoor, S. Majumdar, B. Nandy, "Dynamic Allocation of Sensor Nodes in Wireless Sensor Networks Hosting Multiple Applications", *Proc. 2013 International Symposium on Performance Evaluation of Computer and Telecommunication Systems (SPECTS)*, Toronto (Canada), July 2013.

N.K. Kapoor, S. Majumdar, B. Nandy, "Techniques for Allocation of Sensors in Shared Wireless Sensor Networks", *Journal of Networks*, Vol. 10, Issue 1, January 2015, pp. 15–28.

D.T. Lau, J. Liu, S. Majumdar, B. Nandy, M. St-Hilaire, C.S. Yang, "A Cloud-Based Approach for Smart Facilities Management", *Proc. of the 2013 IEEE Conference on Prognostics and Health Management (PHM)*, Gaithersburg, USA, June 2013.

J. Liu, Y. Xiao, CL.P. Chen, "Authentication and Access Control in the Internet of Things", *Proc. of the 32nd International Conference on Distributed Computing Systems Workshops*, Macao, China, June 2012.

P. Madsen, "Authentication in the IoT – Challenges and Opportunities", *SecureID News*, January 2015, http://www.secureidnews.com/news-item/authentication-in-the-iot-challenges-and-opportunities/. [accessed: October 2015].

S. Majumdar, "Cloud-Based Smart Facilities Management", *Internet of Things: Principles and Paradigms*, Rajkumar Buyya, Amir Vahid Destjardi, Eds., Elsevier, 2016.

S. Majumdar, M. Asif, J.O. Melendez, R. Kanagasundaram, D.T. Lau, B. Nandy, M. Zaman, P. Srivastava, N. Goel, "Middleware Architecture for Sensor-Based Bridge Infrastructure Management", *Proc. of the 15th Communications and Networking Symposium*, Boston, USA, March 2012.

A. McGregor, D. Bennett, S. Majumdar, B. Nandy, J.O. Melendez, M. St-Hilaire, D.T. Lau, J. Liu. "A Cloud-Based Platform for Supporting Research Collaboration", *Proc. of the 8th IEEE International Conference on Cloud Computing (CLOUD)*, New York, June 2015.

L. Ren-Shiou, P. Sinha, C.E. Koksal, "Joint Energy Management and Resource Allocation in Rechargeable Sensor Networks", *Proc. of 29th IEEE Conference on Communications (INFOCOM 2010)*, San Diego, CA, USA, March 2010, pp. 1–9.

J. Tan, G. Tong, "Dynamic Resource Allocation for Target Tracking in Robotic Sensor Networks", *Proc. of IEEE International Conference on Systems, Man and Cybernetics (SMS 2007)*, Montreal, QC, Canada, October 2007, pp. 2354–2359.

C.K. Tham, R. Buyya, "Sensor Grid: Integrating Sensor Networks and Grid Computing", *Invited Paper in CSI Communications, Special Issue on Grid Computing*, India, July 2005.

W3C Semantic Web Standards, "Resource Description Framework (RDF)", February 25, 2014.

11

Summary

The contents of the book are disseminated through 11 chapters. The chapters are divided into two parts: resource management on distributed computing systems, comprising Chapters 2–5 and resource management on distributed data-intensive systems that comprise Chapters 6–10. Chapter 1 provides an introduction to the book and Chapter 11 summarizes the material provided in the various chapters. A summary of the discussion presented in each chapter is described next.

11.1 Chapter Entitled Introduction

This chapter presents some basic information about the different types of parallel and distributed systems discussed in this book.

11.1.1 Types of Computing Environments

Three types of computing environments are defined: single CPU-based multiprogrammed systems, shared memory multiprocessor systems, and distributed memory systems. In the single CPU-based systems, only a single CPU is used. By incorporating multiprogramming into the system multiple processes are allowed to run on the system concurrently. Multiple CPUs are deployed in shared memory multiprocessor systems. Each processor can run one or more processes. The processes can interchange information with the help of shared variables that can be stored in shared memory and accessed by any of the processors. In a distributed memory system, multiple CPUs are provided. But each CPU can only access its local memory directly. Since no shared memory is available, processes running on different CPUs communicate with each other through message passing.

Resource Management on Distributed Systems: Principles and Techniques,
First Edition. Shikharesh Majumdar.

11.1.2 Units of Computation: Processes and Threads

An application deployed on the system can comprise multiple units of computations called processes and threads. A process (also referred to as a heavyweight process) is characterized by the application code that it executes and the memory it uses as well as a state. The memory is used to store the variables used by the process and its stack. The state of a process describes what it is doing at a given point in time. Process states include executing, blocked, ready to run, and suspended. A description of these states and when a process transitions from one state to another are described. A process may comprise multiple threads. A thread (also referred to as a lightweight process) has its own stack, a program counter, and a register set. It typically shares the global data set and code with the other sibling threads in the same address space corresponding to the parent (heavyweight) process.

11.1.3 Resource Management

A computing environment comprises various resources such as processors, memory, and storage devices. Effective management of these resources is essential for effectively utilizing the existing resource pool in the system. Three important operations need to be performed in the context of resource management on a parallel/distributed system. These include the following.

- Resource Provisioning: is used to determine the number of resources of each type required for supporting the given workload and the quality of service requirements.
- Resource Allocation: concerns which unit of work running on the system should be allocated to a given resource instance.
- Resource Scheduling: focuses on the order of execution of work units that are mapped on a given resource during resource allocation.

Five principles underlying the devising of resource management algorithms are described in the book.

- *Principle 1: Use Knowledge of Application/Workload Characteristics:* refers to the use of knowledge of the resource demands of the various applications running on the system that is often used in designing effective resource management algorithms.
- *Principle 2: Monitor and Adjust:* refers to the measurement of system and application state at run time and making adjustments to allocation of resources and scheduling of the units of work in applications to improve system performance.
- *Principle 3: Use Knowledge of System Characteristics:* refers to the use of characteristics of the resources themselves for performing effective allocation and scheduling.

- *Principle 4: Load Balancing:* is a well-known performance optimization technique that advocates the balancing of the workload executing on the various resources so that system performance is enhanced.
- *Principle 5: Static versus Dynamic:* refers to the type of resource management algorithms. For static algorithms resource allocation and schedules are static and do not change with time. Dynamic algorithms, on the other hand, can react to changes in system state and change the allocation of resources and schedules at run time if deemed necessary by the resource manager. Depending on the type of the system and user requirements a static or dynamic resource management algorithm is to be used.

Various resource management algorithms that use these principles are discussed in the later chapters of the book.

11.1.4 Various Forms of Distributed Systems

Distributed systems started with a set of loosely coupled nodes communicating via RPC and have undergone various steps of evolution. Heterogeneity in implementation technology used at the various nodes was addressed through the introduction of middleware such as CORBA and DCOM. For example, a CORBA-based system supports the cooperation between two applications irrespective of the programming languages used in the implementation of the applications as well as the operating systems on top of which the applications are run. Parallel and cluster computing were invented to enhance system performance by using multiple computing elements on the system.

More recently, grids and clouds have been receiving attention from system users and researchers. By using virtualization at different levels, these systems can provide a range of varied types of resources on demand to users. Distributed computing is also being divided into multiple levels. Edge computing, for example, advocates performing computations at two levels: near the edge as well as on backend systems such as a cloud. Some of the computations, the results of which are to be made available in a short period of time, are performed on the edge devices, whereas computation-intensive components that do not have low latency requirements are performed on a more powerful backend system such as a cloud. Management of resources on grids and clouds forms an important component of the book spanning multiple chapters.

The advancement in sensor technology that has led to ready availability of low-cost sensors is fueling work on sensor-based smart facilities that are often referred to as cyber-physical systems. Examples include sensor-based bridges, buildings, and smart industrial machinery such as aerospace engines. Management of resources on such hybrid systems that deploy computing, storage as well as sensor nodes is also included in the book.

11.2 Chapter Entitled Characterization of Parallelism in Applications

This chapter discusses attributes of parallel/distributed applications that have an intimate relationship with system behavior and performance. Two types of characteristics are described. The first type includes the attributes that can be captured through graphs. Examples are described in the following section.

11.2.1 Graph-Based Characteristics

The following discussion focuses on the modeling of computations and their characteristics.

Precedence graph: captures the dependencies (precedence relationships) among the various work units (e.g. processes and threads) that need to be satisfied for the correct execution of the application. A precedence graph is a directed acyclic graph in which each work unit is modeled by a vertex and the precedence relationships between the vertices are expressed through directed edges.

Parallelism Profile: is a plot of the number of work units that can execute simultaneously on the system versus time. This characteristic provides the number of work units that can execute simultaneously at a given point in time when there is no dearth of processing resources.

Shape: captures the relationship between the proportion of the application execution times and the degree of parallelism, that is, the number of work units that are active simultaneously when there is no dearth of processing resources.

Completion Time – Efficiency Profile: shows the relationship between the completion time and the processing efficiency achieved for various numbers of processors allocated to the application. The X-axis shows the time of completion for a given number of processors, whereas the Y axis shows the corresponding efficiency.

11.2.2 Single-Point Characteristics

The graph-based characteristics discussed earlier are fairly detailed. Single-point characteristics are often used when the overhead of using such detailed characteristics is too high for a given use case. The single-point characteristics that were introduced in the chapter include average parallelism, fraction of sequential work, maximum parallelism, and the knee of the application time efficiency profile. They aptly capture the system behavior when the application runs on the system and has a strong relationship with the system performance achieved.

11.2.3 Performance Measures

Performance metrics such as application completion time, job turnaround time, and utilization of resources are well-known. The chapter introduces the reader to two additional performance metrics that are important in the context of parallel applications comprising multiple units of work: speedup and efficiency. The first captures the improvement in system performance resulting from parallelizing the application and running it on multiple processing resources, whereas the second is related to how efficiently these processing resources are used when the application runs on the system. The relationship between these performance measures and application characteristics is discussed in the chapter.

11.3 Chapter Entitled Resource Management Techniques for Distributed Computing Systems

This chapter focuses on resource management in distributed systems, including both shared memory and distributed memory systems running applications that may comprise multiple threads/processes. Both resource allocation, which refers to the mapping of a given work unit (process/thread) to a processing resource and scheduling, which refers to the determination of the order in which the different work units mapped on a given resource are executed, are considered.

The chapter starts with an example that shows the complexity associated with the allocation problem and describes Graham's anomaly, which shows the possibility of anomalous system behavior that can occur in the system.

Finding an optimal solution for resource management is known to be an NP-Hard problem in general. However, for certain classes of applications, an algorithm such as Hu's algorithm that finds the optimal solution in polynomial time is described. Both the algorithm and its limitations are discussed. The chapter provides a discussion of heuristic algorithms that can be applicable to more general classes of applications. Examples of such algorithms such as Largest Processing Time First and Coffman's Multifit are discussed. Three general principles for the devising of new heuristic resource allocation techniques are explained with the help of examples.

The discussion on resource allocation is followed by a discussion of scheduling techniques for grids and clouds. Two types of user requests for scheduling that include on-demand (OD) requests and advance reservation (AR) requests are described. ODs are to be satisfied on a best-effort basis, whereas a service level agreement (SLA) that includes a deadline needs to be satisfied when scheduling AR requests. A scheduling algorithm called *Scaling through Subset Scheduling* is described. The algorithm can be used on systems subjected to a workload comprising both ODs and ARs. Each AR in such a workload is characterized by the earliest start time, a user-estimated execution time and a deadline by which the request must be completed. A simulation-based performance analysis is presented. The simulation results capture the relationships among various performance metrics and the different system and workload parameters.

SSS focuses on scheduling a single resource. An algorithm call *Grid Scheduling with Deadlines* (GSD) that can handle requests with SLAs requiring multiple resources on a grid or cloud is described next. ARs on the system are allowed to hold multiple resources at the same time. GSD has two components: a job ordering heuristic and a node ordering heuristic. The job ordering heuristic is used to determine the order in which the requests (jobs) that arrived on the system are going to be executed. The node ordering heuristic is used to determine which of the resources (nodes) will be used for serving a giving request.

In addition to scheduling on grids, and clouds scheduling on client–server systems is also discussed in this chapter. In such a system, one or more clients can request service from a common pool of servers, each of which can provide a different service. Examples of such servers include file servers, database servers, and print servers. The client sends requests to a server through a network. The server performs the desired operation and replies back to the client with the results of the operation. A number of scheduling algorithms such as First Come First Served, Shortest Request First, and Longest Path First are described. The first does not use any knowledge of request characteristics that are used in the two other algorithms to determine which service request will be executed next.

11.4 Chapter Entitled Resource Management on Systems Subjected to Uncertainties Associated with Workload and System Parameters

On real grids and clouds there are often uncertainties associated with the characteristics of the system and/or workload. Robust resource management techniques should be able to handle such uncertainties. Techniques for handling two types of uncertainties are discussed:

- Handing errors associated with user estimates of task execution times.
- Handling lack of *a priori* knowledge of local operating systems on individual nodes in a large system such as a grid or cloud.

11.4.1 Handing Errors Associated with User Estimates of Task Execution Times

Resource management algorithms discussed in Chapter 3 are often based on task execution times. It is difficult to get *a priori* knowledge of the exact task execution times. Users provide estimates of the execution times of tasks that constitute the workload the system is subjected to. Such estimates are often error-prone. The detriments in system performance that can occur in the system are discussed in this chapter. Overestimation of task execution times can give rise to premature job terminations, leading to a deterioration in resource utilization. Underestimation of task execution times can lead to job abortions that can negatively affect system performance.

Schedule Exceptions Manager, SEM, and the *Prescheduling Engine*, PE, can mitigate the diverse effects of the error associated with user estimates of task execution times. Both of these systems are described in the chapter. SEM executes at run time and tries to dynamically adjust the existing resource schedule to handle a premature job termination resulting from an overestimation of task execution times and an unexpected increase in job execution time resulting from an underestimation of task execution times. PE is used before job execution. The prescheduling engine's objective is to tackle the problem of potential overestimation of execution times by overbooking a resource and artificially reducing the estimates of job execution times provided by the user before resource allocation and performing the respective schedulability analysis. A simulation-based performance analysis is described to demonstrate the effectiveness of these systems.

The effectiveness of these error-handling mechanisms is discussed using a simulation-based analysis. The results demonstrate the efficacy of SEM and PE and provide insights into system behavior and performance.

11.4.2 Handling Uncertainties Associated with the Local Scheduling Policy

In addition to the uncertainties discussed earlier, another type of uncertainty that needs to be dealt with stems from the heterogeneity of resources in distributed systems, including clouds. A cloud data center can use hundreds of hardware servers. These may have been purchased at different times. Resource availability can also be dynamic. Resources may be borrowed temporarily from another distributed system. As a result, the local operating system used by the resources can vary from one resource to another. It is not appropriate to assume that the cloud resource manager will have exact *a priori* knowledge of the local operating systems running at all the resources. Techniques for building robustness in the cloud resource manager to mitigate this uncertainty are described in this chapter to address this problem.

The resource management techniques discussed in this chapter focus on systems comprising two types of resources: transparent resources for which the details of the scheduling technique used by the local operating system are known and opaque resources for which the local resource management policies are unknown. Any schedulability (AS) criterion is introduced to handle opaque resources. Requests that are to be handled by opaque resources are scheduled using the AS criterion so that it can be ensured that the request SLAs are met. The requests that are to be handled by transparent resources are scheduled by using techniques that use the knowledge of the respective local scheduling policies used at the resources. The usefulness of using hybrid techniques that use both of these scheduling techniques is discussed with the help of a simulation-based analysis. The discussion shows that the AS-based criterion can effectively improve the acceptance rate of requests submitted to the system that leads to an increase in the revenue rate for the service provider.

11.5 Chapter Entitled Resource Auto-Scaling

The workload a system is subjected to often changes with time. If the system is provisioned for adequately meeting the peak workload demand, then some of the resources will remain unutilized during periods of time during which the workload intensity is lower. If the system is provisioned for meeting the lowest workload demand, then system outages will occur when the system is processing a higher intensity workload. On the other hand, if the system is provisioned using the average workload demand, then there will be outages during periods of time when the demand is higher than average and resources will remain unutilized during periods of time when the system is subjected to a lower workload intensity. Thus, to handle all of these situations, the number of resources provisioned on the system needs to change dynamically. This chapter discusses resource auto-scaling, which refers to changing the number of resources in the cloud with time. The difference between vertical auto-scaling and horizontal auto-scaling is discussed first. The remaining chapter focuses on the horizontal auto-scaling of CPU resources.

The main goal of an auto-scaling technique is to adjust the number of resources in such a way that resource wastage is minimized while ensuring that client SLAs are met. This ensures a reduced cost of operation and the concomitant increase in profit for the service provider as well as user satisfaction resulting from providing an adequate quality of service.

The chapter describes the system and background terminology, including the public cloud provider that contains the resources, a single client enterprise that sends requests for resources, and an intermediary cloud provider that resides between the public cloud provider and the single client enterprise to provide a virtual private cloud (VPC) and to perform auto-scaling on the VPC. The workload generated comprises both OD requests that are to be satisfied on a best-effort basis and as well as AR requests that are associated with the respective SLAs that include the earliest start time, job execution time, and a deadline by which the request must be completed.

Three types of auto-scaling techniques are described: reactive auto-scaling, proactive auto-scaling, and hybrid auto-scaling. For *reactive auto-scaling*, resources are added or removed from the VPC after a predefined change in the workload has occurred, whereas with *proactive auto-scaling* resources are added or removed from the VPC proactively in advance before the actual change in workload has occurred. The proactive auto-scaling technique described tries to predict the future workload and increase/decrease the number of resources in anticipation that a certain predicted change in workload is about to occur. *Hybrid auto-scaling* combines both the reactive and proactive auto-scaling approaches and comprises two components: a reactive auto-scaling subsystem and a proactive auto-scaling subsystem. The reactive auto-scaling subsystem uses a metric called blocking ratio (B), which is the ratio of the number of requests that have been rejected and the total number of requests that have arrived on the system. The client provides a desired value of B called B_{spec}. The reactive auto-scaler checks the value of B on the system when each request arrives and makes auto-scaling decisions in such a way that B does not exceed B_{spec}. After ensuring that this condition is satisfied, the reactive auto-scaler tries to maximize the profit for the intermediary cloud provider and adjusts the number of resources accordingly. Two metrics: the profit earned and the cost borne by the user are used to determine the efficacy of the algorithm described in this chapter. A simulation-based performance analysis shows that the algorithm leads to broker profit and a lower user cost in comparison to a system that does not use auto-scaling.

With proactive auto-scaling, a decision regarding changing resource allocation is made in advance, before the change in system state actually occurs. The decision is applied, however, at a predicted point in time. The proactive auto-scaling algorithm described in this chapter attempts to predict a future system state and decides if a change in the number of resources in the VPC is to be made at a given future point in time. The auto-scaler predicts the future workload by using a machine learning technique and decides whether or not additional resources are to be acquired in the future to handle this change in workload. The workload predictor in the auto-scaling system is often used with a performance model to determine the number of resources required to meet the specified grade of service when the system gets subjected to the predicted workload. If additional resources are required, the system determines if acquiring the resources at the predicted time will lead to a profit for the intermediary cloud provider. If there is a profit the system decides to acquire the resources.

A simulation-based performance analysis of the proactive auto-scaling techniques is presented. The performance results show that both profit for the intermediary cloud provider and a lower user cost in comparison to a system that does not use auto-scaling can accrue from the use of proactive auto-scaling.

The hybrid auto-scaler combines reactive auto-scaling with proactive auto-scaling. The hybrid auto-scaler runs two threads, one for reactive auto-scaling and one for proactive auto-scaling. The reactive auto-scaling algorithm runs on every request arrival. It decides to acquire a new resource if: (i) no currently available resource on the system satisfies the SLA of the newly arrived request and the acquisition of the resource will lead to a profit for the broker or (ii) rejecting the request will lead to a B that is higher than the user-specified value of B_{spec}. The proactive auto-scaling thread is run periodically and predicts the characteristics of the next few requests to arrive on the system and simulates the resource management algorithms to check if additional resources are to be acquired in the future. It decides to acquire resources if (i) no available resources satisfy the SLA of one or more of the upcoming predicted requests and the respective resource acquisitions lead to the prediction of a profit for the broker or (ii) rejecting some of the upcoming predicted requests leads to a B that is higher than the user-specified value for B_{spec}. A simulation-based performance analysis is presented to demonstrate the efficacy of the hybrid auto-scaling approach.

11.6 Chapter Entitled Resource Management on Systems Running MapReduce Jobs

This chapter concerns clouds and clusters that run data and compute-intensive applications that use the MapReduce framework for parallel processing. With MapReduce, the input data is split into a number of chunks, each of which is processed by a map task. Multiple map tasks run on the system thereby concurrently processing different chunks of the input data. The results produced by the map tasks undergo a "shuffle and sort" operation, after which they are processed by multiple reduce tasks that combine the results produced by the map tasks.

The chapter explains the MapReduce model with the help of examples. Resource management techniques for MapReduce jobs are discussed next. A number of existing resource managers available for the well-known HadoopMapReduce system are described. These include a FIFO, a Fair Share, and a Capacity scheduler. These algorithms schedule jobs on a best-effort basis. Resource managers for systems running MapReduce jobs with SLAs are described next. Each job is associated with an SLA that includes the earliest job start time, task execution time, and a deadline by which the job must be completed. A budget-based resource manager for aptly handling MapReduce jobs with SLAs is then described. The algorithm and a simulation-based performance analysis are presented. The results of the performance analysis show the effect of system and workload parameters on system performance as captured by two performance metrics: proportion of late jobs that miss their deadlines and mean job turnaround time.

A novel resource management technique that uses constraint-based programming is introduced. The constraint programming (CP) model is described first. The resource management algorithm that uses the CP model is presented next. It takes the SLAs associated with the MapReduce jobs, generates a mapping of the map, and reduce tasks on the resources so that the SLAs for all the jobs accepted to run on the system are satisfied. The performance of a prototype implementation of the resource management algorithm on a Hadoop cluster is analyzed. Two performance metrics: the proportion of late jobs and the mean job turnaround time, were measured on the system. A comparison with the well-known Earliest Deadline First (EDF) scheduler demonstrates the superiority of the CP-based technique over EDF.

Estimation of task execution times is hard and there is often an error/inaccuracy associated with the execution times of tasks estimated by users that are sent as part of the SLA. Overestimation and underestimation of task execution times performed by the user can lead to a deterioration in system performance. When a task execution time is overestimated, the task completes prematurely before the time that has been determined by the resource manager. This tends to introduce idle times on a resource, lowering its utilization. When the execution time of a task is underestimated, the task remains incomplete at the expected completion time computed by the resource manager, leading to a potential abortion of the job. Such an abortion tends to reduce the useful utilization of the resource. Building robustness into resource management techniques for handling such inaccuracies in user estimation of execution times of map and reduce tasks is thus important for clusters and clouds subjected to a MapReduce workload. Two techniques for handling errors associated with user-predicted task execution times that can be incorporated into the constraint programming-based resource manager mentioned in the previous paragraph are described. One of them, called the prescheduling error handling technique, is used before running the tasks to adjust the user-supplied task execution times before sending the jobs to the resource manager. The resource manager keeps track of both the estimated and actual run time of jobs that have already completed on the system. The difference between these two values is used to compute an error factor that quantifies the degree of inaccuracy in estimates of task execution times provided by the user. The average error factor computed from such statistics for all jobs that have completed on the system is used to adjust the task execution time specified in future submissions by the user.

Another error handling technique, called the runtime error handling technique, is used after the job has started execution and attempts to mitigate the adverse effects of overestimation and underestimation of task execution times. The algorithm is invoked in two situations: (i) when task execution time was overestimated and the task was completed before its scheduled end time and (2) when the task execution time was underestimated and the task was not completed by its scheduled end time. Mechanisms for handling both of these situations are described.

A performance analysis of the prescheduling and the runtime error handling techniques demonstrates the importance of using error handing to improve system performance.

11.7 Chapter Entitled Energy-Aware Resource Management

This chapter focuses on energy-aware resource management on systems that process MapReduce jobs with SLAs. Techniques that target the meeting of SLA requirements and reducing the energy consumption by the system at the same time are described.

The total aggregate energy cost of a data center is a significant component of its monthly operational cost. Reducing power consumption through energy-aware resource management is the focus of this chapter. Dynamic Frequency and Voltage Scaling (DVFS) is a technique that can be used for controlling the power consumed by the server. This technique leverages the impact of clock frequency on the power consumed by a CPU. Reducing the clock frequency reduces the CPU operating frequency. Although this leads to a decrease in the speed of instruction execution it decreases the power consumed by the device. Another technique used for reducing server energy consumption is consolidating several executing tasks on a subset of the total number of servers and powering down the idle servers created by this consolidation operation. Once more capacity is needed due to the arrival of a new workload, one or more of these idle servers are woken up and the existing tasks are redistributed among these active servers on the system if needed. The chapter discusses both of these techniques.

The equations related to the relationship between energy and CPU operating frequency are fundamental for the DVFS-based technique and are presented first. The energy-aware resource management algorithm is presented next. It uses a constraint programming-based approach that is similar to the one discussed in the context of resource management on systems processing Map Reduce jobs with SLAs that was presented in an earlier chapter. One of the additional objectives in this constraint programming model is the minimization of the energy consumed by the servers. Energy savings come at the cost of slowing down the CPU operations, which may lead to an increase in the execution time of a task. The algorithm adjusts the CPU frequency in such a way that the job SLAs are not violated, but the server energy consumption is reduced. Simulation-based performance analysis shows the relationship between various workload and system parameters (e.g. job arrival rate and the number of resources used in the MapReduce cluster) and the performance of the DVFS-based resource management technique.

A high-level description of the second technique that attempts to reduce the number of active servers for energy conservation is described next. The technique classifies servers into two classes based on their CPU load: active and nonactive. A server's CPU load, which is higher than a predetermined threshold, is deemed active; it is classified as nonactive otherwise. Tasks on the nonactive server that have not yet started execution are transferred to an active server. When the last task executing on this server is completed, the server is deactivated. If such a server cannot be found without missing deadlines of some accepted jobs, then the task transfer is not performed. An active server receiving a transferred task is selected in such a way that the deadline constraints of tasks are not violated. When a newly arrived task cannot be allocated with the current number of active servers, one of the deactivated servers is activated again.

11.8 Chapter Entitled Streaming Data and Complex Event Processing

This chapter considers streaming systems in which streams of data tuples are processed.

Examples of streaming systems include social networks and sensor-based systems monitoring physical phenomena in real time. The chapter focuses on two issues related to stream processing: resource management for stream processing systems for priority-driven scheduling and complex event processing. The workload on the systems discussed is dependent on the contents of the data that arrive on the system.

11.8.1 Resource Management

Three popular stream processing platforms: Apache Spark, Apache Storm, and Apache Siddhi are described first to provide the background for the following discussion on stream processing performed on these platforms.

Streams comprise tuples of data originating from a given source such as a social network and sensors in an IoT-based system. Some of these tuples may need to be handled on a priority basis and processed prior to other tuples that may have reached the system earlier. The tuple priorities often depend on the tuple contents. The dynamic data-driven scheduler (DDPS) for priority-based scheduling discussed in the chapter adjusts the priority of data elements based on the existence of high-priority keywords in the data stream. The algorithmic components of the scheduling technique are discussed first. The DDPS technique can be used in conjunction with the Spark Streaming system: a component of the well-known Spark data processing platform. The performance of the DDPS technique deployed on an Apache Spark platform is discussed next.

Priority-based resource management techniques for multi-tenant storm clusters are described. The resource manager on such a system needs to distribute the resources among multiple storm topologies based on their respective priorities. The well-known isolation scheduler that can isolate the resources allocated to the different applications from one another but cannot handle priority scheduling is described first. Two priority-based schedulers are described next. For the static priority scheduler the priorities of the topologies are fixed and are specified before run time. In the case of dynamic priority schedulers, a topology priority can change dynamically at run time. The priority of a topology is changed upon receiving a "trigger ON" signal by the priority manager. Both an algorithmic description and performance analysis of each of these static and dynamic priority-based resource management techniques are presented.

11.8.2 Complex Event Processing

The detection and management of complex events are discussed in the second part of this chapter. A complex event is created by multiple low-level raw events. A certain pattern of raw events is required to occur for producing a complex event. The architecture of a complex event processing (CEP) system is described. The functioning of a CEP is described with the help of a remote patient monitoring (RPM) system. A chronic cardiac patient is fitted with body sensors that monitor various statistics associated with the patient such as their heart rate and respiration rate. The sensor outputs generate raw evens streams that are analyzed by a CEP engine. The engine tries to find patterns that predict impending health problems for this patient.

Two versions of the RPM system are discussed. In the server CEP, SCEP, system, the sensors communicate through Bluetooth or Wi-Fi with the gateway agent, which is a mobile device (MD) such as a smartphone. This gateway agent runs a Server Event Processing application that receives the sensor streams and sends them to a Hospital Server (IHS) that detects complex events from the sensor streams received from MD. The results of this complex event detection are then displayed on a dashboard. In the Mobile CEP, MCEP, version, the complex event processing is performed on the MD itself. Communication between the MD and IHS occurs only when a complex event is detected. The implementation of prototypes for each of these systems is described. The important features of both the SCEP and the MCEP systems are discussed. This discussion provides insights into the relative merits and demerits of each system. A performance analysis shows that lower energy consumption and lower latency are achieved by the MCEP system in comparison to the SCEP system.

11.9 Chapter Entitled Data Indexing and Filtering Techniques for Big Data Systems

This chapter concerns big data systems. It first describes a number of important characteristics of big data: volume, velocity, variety, and veracity that are responsible for the increased complexity of big data processing. Two different approaches for addressing performance challenges resulting from these characteristics are: data indexing and reducing data volume through filtering of data. A discussion of both of these approaches is presented in the chapter.

11.9.1 Data Indexing

Indexing is a technique for reducing the latency of performing searches on a large document set. An index enables a quick location of the set of documents that contain one or more specific keywords. Note that all the words in a document are not indexed. Only a specific set of words identified in a keyword extraction step are associated with an index.

Both exclusive keyword search queries and binary search queries are described in the chapter. Exclusive keyword search queries involve only single keywords. In addition to exclusive keyword search queries Boolean queries are often used in information retrieval. A Boolean query comprises multiple keywords that are connected by one or more Boolean operators such as OR, AND and NOT. Of particular interest is the Boolean AND query that is often used in many search engines such as Google. Single keyword queries are often not very useful in information retrieval because they may result in a large set of documents with many items in the list being of no interest to the user. A Boolean query, on the other hand, can be used to narrow the search. Boolean AND queries are the focus of attention of the discussion on data indexing presented in this chapter.

Two indexing techniques called inverted indexing and graph-based indexing that are useful in information retrieval are described. An inverted index is a data structure that maps a keyword to a list of documents, which is also referred to as a postings list. The graph-based indexing technique, GBI, uses a directed graph to index the keywords appearing in the documents. Such a graph comprises a set of nodes connected by directed edges. The label of each node is one of the keywords to be indexed. GBI is very useful in identifying documents containing multiple keywords and is thus used in the context of AND search queries.

A simulation-based approach is used in analyzing the performance of the different algorithms described in the chapter. Both the search latency observed with the indexed data and the respective indexing overheads are discussed.

11.9.2 Data Filtering

Data filtering techniques that can reduce the volume of raw data were described. Techniques for handling both stored data and streaming data were discussed. The described algorithms filter raw data by using user-provided keywords/sentences. Enhanced filtering algorithms that can handle multiple user-specified keywords/sentences connected by Boolean operators such as AND, OR, and NOT were discussed.

Filtering techniques using machine learning are CPU-intensive processes and parallel processing techniques are used to reduce the filtering time. Proof of concept prototypes deployed on Apache Spark that help reduce the data filtering time were described. System performance observed with different combinations of systems and workload parameters were discussed. Insights into system behavior and performance that result from the performance analysis were reported. The efficacy of data filtering is captured through a newly introduced metric called filtering efficiency, which is the ratio of the search time achieved with raw data and that achieved with filtered data. Filtering efficiencies that are significantly higher than 1 are reported in a number of experiments, thus demonstrating the effectiveness of the data filtering techniques.

11.10 Chapter Entitled Sensor-Based Systems

Managing systems comprising both sensing and processing elements that are often used in the context of a smart environment is discussed in this chapter. Examples of such smart environments range from sensor-equipped bridges to sensor-based industrial machinery such as airplane engines. Irrespective of the exact nature of the smart environment a set of common steps are performed for their monitoring and management. These include sensing of the respective phenomena of interest and collecting the sensor data, analyzing the collected sensor data; and making decisions regarding the management/maintenance of the facility.

Although sensors are typically characterized by limited processing capability the sensor-based smart system can utilize the computing resources in a cloud for processing sensor data as well as its storage. Such processing may be required for monitoring the state of the system as well as for controlling its operations. The processing can be performed both by an edge cloud and a remote backend cloud.

11.10.1 Smart Facilities Management

The first part of the chapter concerns sensor-based smart facility management. The architecture for a cloud-based smart facility is presented and a common set of varied middleware services that are useful in the context of various smart environments is described. These services range from a user authentication service to a resource management service. This is followed by a discussion of two examples of smart facilities. The first is the middleware required for sensor-based bridges and the second concerns a cloud-based platform for supporting collaboration among participants of a team of researchers.

11.10.2 Management of Wireless Sensor Networks

The second part of this chapter focuses on resource management on wireless sensor networks (WSNs) that comprise a cluster of sensors inter-connected by a wireless network interfaced with a cloud or grid through a proxy node. The WSN is shared by multiple applications. The applications send their requests to the proxy node that runs a resource management algorithm to allocate sensors to applications as well as schedule the application requests destined for the same sensor. Both sensor scheduling and sensor allocation are described.

11.10.2.1 Scheduling of Requests

At any point in time multiple applications can request for a specific sensor. The job of the scheduling algorithm running at the proxy is to decide the order in which these requests will run on the given sensor such that high system performance is achieved. The performance metric used for analyzing the performance of the sensor scheduling algorithms is the overall mean response time. Two classes of scheduling algorithms are discussed in this chapter: one that does not use any information about network/system and application characteristics and one that is based on application and network characteristics. The algorithms in the second class are further divided into three categories based on the type of knowledge about the network or application they use: knowledge of the application characteristics only, knowledge of the network characteristics only, as well as knowledge of both application and network characteristics. After describing the various algorithms in different classes, a report on a simulation-based performance analysis is presented. The simulation results demonstrate the utility of using the different types of knowledge about the applications and the network in scheduling. The impact of various workload characteristics such as the request arrival rate, the data size, and the number of applications sharing the WSN are discussed.

11.10.2.2 Sensor Allocation

Applications send requests for the sensors in the WSN. The sensor allocator technique determines which sensors in the WSN are to be allocated to which application. Two classes of allocation algorithms are discussed in the chapter: *static* and *dynamic*. For static allocation algorithms, sensor allocation to applications is performed only once and does not change during the lifetime of the application, whereas with dynamic allocation algorithms, the mapping of sensors to the application generating a request is performed dynamically upon request arrival.

Each of these classes of algorithms can be divided into two sub-categories: knowledge-based and knowledge-free. Two types of knowledge-based static algorithms are described. In one, the execution time of application requests is used, whereas the other uses the size of response messages. The knowledge-free static algorithms do not use any knowledge of the request characteristics or the network. The knowledge-based dynamic algorithms use one or more characteristics of the network and the application requests, whereas for the knowledge-free algorithms no knowledge of the application or network/system characteristics is used in determining which sensor will be allocated to which application. The knowledge-based dynamic algorithm can again be classified into three classes: one that uses the knowledge of the network, one that uses the knowledge of the application characteristics, and the other that uses the knowledge of both the network and the application.

Performance of the allocation algorithms is compared using simulation experiments. The effect of various workloads and system parameters on two performance metrics, minimum energy (the energy of a sensor node that has the least amount of remnant energy at the end of the simulation period) and network lifetime are described.

Index

Resource Management on Distributed Systems: Principles and Techniques, First Edition. Shikharesh Majumdar.

Printed and bound by CPI Group (UK) Ltd, Croydon, CR0 4YY

17/09/2026

14964255-0001